T4-ALD-963

# Environment, Energy, Public Policy: Toward a Rational Future

# Environment, Energy, Public Policy: Toward a Rational Future

Edited by
**Regina S. Axelrod**
Adelphi University

**LexingtonBooks**
D.C. Heath and Company
Lexington, Massachusetts
Toronto

**Library of Congress Cataloging in Publication Data**

Main entry under title:

Environment, energy, public policy.

1. Energy policy–United States–Addresses, essays, lectures. 2. Environmental policy–United States–Addresses, essays, lectures. I. Axelrod, Regina S.
HD9502.U52E462 333.79'0973 79-3523
ISBN 0-669-03460-6



Published simultaneously in Canada

Printed in the United States of America

International Standard Book Number: 0-669-03460-6

Library of Congress Catalog Card Number: 79-3523

*For Gregg*

# Contents

# Preface

This book developed as a result of a conference held at Adelphi University, June 8-9, 1979. Energy and environmental problems have been most often studied from diverse and disparate perspectives, and I believed that some serious attempts should be made to provide an interdisciplinary and interactive framework for exploring these issues. The authors of these chapters reflect the wide-range of participants at the conference.

The purpose of the book is to draw attention to the complexities of energy and environmental problems and to demonstrate the relationship between authoritative actions and particular aspects of this issue. The chapters all reflect on the convergence of energy and environmental policy from either an institutional or a policy-process perspective.

Regina Axelrod outlines the parameters of the environment-energy dilemma in chapter 1. It is suggested that the dimensions of this very complex issue make it the most challenging to ever confront our political system.

In chapter 2, Dankwart Rustow discusses the current oil crises characterized by high price and unreliable supply, which triggered the immediate concern in energy and the subsequent search for ways to meet energy needs with minimal environmental impact. He dispels some commonly held myths. For example, it is believed by some that the energy crisis resulted from actions of OPEC nations; in actuality the scarcity of petroleum supply would have caused disruptions in the market by the next century anyway. Rustow notes that as the price of oil has increased, so has the cost of alternative fuels, which he suggests indicates the pervasive control of fuel supplies by energy companies. As part of a general solution, Rustow recommends that emphasis be placed on reducing the demand for fuel because the creation of alternative fuels will be costly and will cause environmental problems and thus will not yield immediate relief.

The nature of the representativeness of public opinion on energy and environmental issues is addressed by Helen Ingram and Nancy Laney in chapter 3. The more coherent and consistent voter attitudes are, the easier it is for legislators to respond to an issue. Using data from the American Southwest, the authors examine the intensity, consistency, and distribution of public opinion on specific energy and environmental issues. They look at the dilemma facing state legislators who must choose between responding to their constituents' beliefs or acting on their constituents' behalf. They conclude that the nature of public opinion reflects legislators' ability to strike a balance among competing issues. Lacking support for comprehensive policy alternatives that resolve energy-environmental conflicts, the vague mandates to legislators leave them free to choose symbolic or piecemeal actions rather than substantive ones.

In chapter 4, Michael Kraft focuses on the development of a national energy policy in the Congress. He describes congressional behavior during the stages of agenda setting, formulation of proposals and adoption, and examines the roles of interest groups, public opinion, political climate and executive leadership. Kraft concludes that between 1977 and 1979, Congress acted in a predictable fashion, bypassing comprehensive energy policy making because of political and institutional obstacles. Like Ingram and Laney, he sees a positive role for public opinion if the public becomes informed and can register clear preferences.

Walter Rosenbaum also looks at the federal legislative response to energy and environmental concerns in chapter 5, suggesting that critical decisions are made for specific political reasons that have little to do with content. Decisions are based on criteria that ignore the desirability of outcomes on substantive grounds. He concurs with Kraft that congressional policy making is less dependent on analysis of policy than on the influence of overpowering political forces. Congressional funding of research and development in energy typically has been influenced by demands from vested interests, both public and private, which have high stakes in particular outcomes of funding decisions. Rosenbaum aptly describes the impact of coalition building and logrolling on project funding and its environmental consequences.

Over the past decade, energy and environmentally related cases have come to play an increasingly significant role in another branch of government, the judiciary. Two major areas of controversy involve the siting of nuclear and fossil-fuel electric-generating plants. Lettie Wenner warns in chapter 6 that the courts are becoming increasingly burdened because of the highly technical nature of the cases and the lack of conclusive information on which to base their decisions. She notes the courts' differential treatment of nuclear and fossil-fuel plant cases; courts are more critical of environmental impairment in fossil-fuel plant cases and less likely to intervene in nuclear plant cases, deferring instead to nuclear industrial and administrative expertise. Wenner surveys court cases, notes trends in judicial attitudes, and speculates on the ability of the courts to render equitable decisions in an increasingly complex area.

In chapter 7, Karen Burstein offers an intimate examination of a specific decision-making situation involving a state public utility commission on the question of building a high-voltage transmission line to bring Canadian hydroelectric power to New York State. She identifies the compelling forces with which a regulatory agency must contend and the administrative environment where other government units compete for attention. This case demonstrates how alternatives are precluded as a result of steps made early in the decision-making process and the importance of prior commitments, biases, and interpretation of events to fit preconceived notions. Burstein reflects on the narrow scope of the decision-making process and why decisions are made based not on merit but because of particularist and parochial concerns.

David Harrison and Michael Shapiro assess the role of local government in energy management in chapter 8. Although they recognize that politics can impede innovation, they point out that many opportunities remain for new approaches. Since local governments are best suited to determine their specific needs, they can develop unique energy savings plans, such as building and zoning ordinances that reduce energy usage. The authors compare available options that local governments could initiate and evaluate their usefulness in light of national conservation goals.

Gregory Daneke addresses in chapter 9 the role of administrative agencies in the planning and review of environmental and energy policy. He demonstrates some of the difficulties encountered when energy analysts are confronted with environmental assessment; both energy and environmental decision makers lack a framework for full consideration of the complex interactions of their decisions. He examines those factors that must be an integral part of the decision-making process and suggests means for incorporating environmental values into energy decision making.

# Acknowledgments

Organizing a national conference and assembling its papers have been challenging and rewarding tasks. An idea must receive support and can be nurtured only in the proper atmosphere. I am especially grateful to the Adelphi University administration and my departmental colleagues for their steadfast encouragement. Adelphi University President Timothy Costello was an early supporter of my plans, as were Harry C. Davies, dean of the College of Arts and Sciences, and Clifford T. Stewart, dean of academic affairs. I am also indebted to Burton Eckstein, director of the Office of Sponsored Projects, and Ron Cannova, director of university relations, for their expertise and support. Gerald Heeger, former chairperson of the Department of Political Studies and now dean of the University College, and Mel Albin of the Department of Political Studies provided the special friendship and invaluable advice that only devoted colleagues can offer. The director of the Institute for Suburban Studies, Hugh Wilson, has also been a source of immeasurable encouragement. The Institute for Suburban Studies, a research center and a degree-granting program, allowed graduate students to participate in the planning of the conference. Special thanks go to Rosalie Simari for her diligence and hard work; I hope she benefited as much from the experience as I did from her assistance.

This material was prepared with the support of Energy Research and Development Administration Grant DE-FG01-79EV10044 and with financial support from the Ford Foundation and the Environmental Protection Agency. I would like to thank Ruth Clusen, Carol Jolly, Shelly Weinstein, and Herbert Fish, the Office of the Environment, Department of Energy; Steven Reznek, Office of Energy, Minerals and Industry, Environmental Protection Agency; and Allan Pulsipher, the Ford Foundation, for their generosity, their confidence, and their imaginative contributions to the success of the conference and this book.

I would also like to express my appreciation to Lenny Kahn, assistant commissioner, New York City Department of General Services; Walter Rosenbaum, University of Florida; Elaine Hussey, California Department of Transportation; Lettie Wenner, University of Illinois; Jon Czarnecki, Office of the Minority Leader, New York State Senate; Gregory Daneke, University of Arizona; and Michael Kraft, University of Wisconsin, for their useful comments and suggestions on the manuscript, and Stuart Nagel for advice on production. Credit for the design and layout of the figures is due to the impeccable work of Angela Perrata. The book could not have been completed without the tireless efforts and expert assistance of Deborah Heineman, administrative assistant of the Institute for Suburban Studies, and Karen Tranchilla, secretary of the Department of Political Studies, whose typing and administrative skills are unsurpassed. Also, Therese Ginty, secretary of the Department of Philosophy, volunteered service when the work load became too heavy.

I would like to take this opportunity to offer a personal expression of gratitude to my former teacher at the Graduate School, City University of New York, Ann Marie Walsh, who has been a constant source of inspiration and intellectual support.

And finally, I am indebted to my parents who have always tried to make the way easier, and taught me the supreme value of the human existence.

# 1 Energy and the Environment: Conflict and Resolution

*Regina S. Axelrod*

One of the most urgent challenges facing the United States is the direction of governmental activity affecting both energy and environmental issues. Although this problem is often presented in an antagonistic context, the future of our civilization ultimately rests on a harmonious relationship between patterns of energy usage and their environmental impact.

Public interest in these issues often has been a reflection of the level of perceived crisis and attention by the media. Interest in the environment and more specifically the impact of technology and industrialization upon human health, plant and animal life, and the irreversible changes in the physical characteristics of the planet reached its peak in the early 1970s. Concern about the nation's ecological future provided a basis for the passage of such legislation as the National Environmental Policy Act (1969), the Clean Air Act of 1970 (and subsequent amendments), the Federal Water Pollution Control Act of 1972, the Clean Water Act of 1977, and the Surface Mining Control and Reclamation Act (1977) and for the creation of environmental agencies on local, state, and federal levels. These actions were followed by a variety of regulatory activities that attempted to shift some of the costs back onto the polluter and encourage the development of more reliable antipollution techniques. However, environmental agencies experienced difficulties in fulfilling their mission; enforcement was never adequately funded and implementation of legislative goals was often left to the states, sometimes resulting in lengthy litigation.

The emergence of an energy crisis now threatens to defeat what little progress has been made. Groups whose immediate agendas are at odds with government activity concerning the environment and wish a change in the order of priorities perceive environmentalists as either too radical (for example, in advocating utopian solutions) or too conservative (by restricting commercial development of wilderness areas, for example). The postponement of deadlines and relaxation of environmental rules reflects pressure from energy and industrial interests, which contend that the stringent application of environmental regulations poses obstacles to energy development and threatens industrial growth and employment. They argue that in order to meet current energy demands, readjustments must be made to previous commitments to protect and clean up the environment. They propose environmental trade-offs without seriously evaluating existing patterns of energy supply and usage or considering the consequences of continued high-level energy use on the stability of delicate ecological systems.

In the past, experts have claimed that maintaining adequate and uninterrupted supplies of energy were necessary to preserve continued economic growth and employment levels and to support prevailing standards of living. Recent evidence has shown that although energy usage decreased in 1979, the gross national product has continued to grow, but at a slower pace. Moreover, it is not clear that an increased use of energy will in fact increase employment.

Eventually conventional energy supplies will become scarce. However, widespread concern over scarcity did not appear until interruptions in the supply of oil occurred, due in part to a controlled market situation in the mid-1970s. This concern was exacerbated by the increasing costs for petroleum products, which squeezed American pocketbooks and led to a flurry of political activity. The American public was faced with the prospect of paying considerably more for petroleum products without apparent alternatives or substitutes. Associated with the widespread impact of higher energy costs were cutbacks in economic development projects and social services supplied by municipal governments at the same time that taxes and unemployment were increasing. Higher prices for all consumer goods, due in part to increased transportation costs, have had a profound economic impact on all Americans.

Scarcity problems aside, the ways in which energy is converted to usable forms affect the future of all ecological systems. Some experts believe that the finite capacity of the environment to assimilate pollutants is quickly approaching. The cumulative effect of a multitude of pollutants originating in our industrial production and energy combustion processes are gradually becoming evident. Increasing concern is being focused on acid rain, especially severe in the northeast United States where high amounts of sulfur oxides in the air have created serious air and water pollution. (Sulfur compounds, a by-product of the fuel combustion process, are highly acidic.) Effects of environmental pollution include increased respiratory illness; increased morbidity and mortality rates; damage to vegetation, animal, and fish life; and dramatic changes in atmospheric composition. Another example is the greenhouse effect. It is believed that carbon-dioxide accumulations from nonstationary sources (such as vehicles) and the increase of combustion of fossil fuels (coal and synthetics) from stationary sources will increase the temperature of the planet, causing climatic changes and partial melting of the polar ice caps. Synthetic fuels (now being given considerable government support) are also likely to produce hazardous waste material, such as sludge and toxic water. Many of the environmental effects of energy production are still unknown. The government is continually identifying new toxic pollutants and has yet to assess their long-range effects. Our standards of safety and risk are still speculative, and research continues as air- and water-quality standards are revised in light of new data.

**Government Response**

The collision between environmental protection and energy needs is manifest in numerous ways. Opposition by state and local authorities and affected constituencies to the dumping of nuclear wastes because of potential hazards and the lack of faith in nuclear technology (exacerbated by the ineptitudes revealed by the Three Mile Island accident) have forced public authorities to reevaluate reliance on nuclear power as a solution to our energy needs. In the past, issues such as nuclear waste disposal were considered engineering problems. Public faith in technology has now eroded. Phrases like "no significant impact" (issued by authorities in response to nuclear accidents) are being challenged, as are plans to utilize nuclear power plants to meet electric demands. The desire by oil companies to drill for petroleum in rich fishing areas such as the Georges Bank off the New England coast, has aroused opposition by fishermen claiming that their livelihood is in jeopardy. The Department of Energy's plans to offset petroleum imports partially through increased use of coal by electric utilities threaten to reverse the improvements made in air quality over the last decade.

**Cost-Benefit Analysis and Regulation**

Generally environmental advocates are now labeled by their opponents in the energy industries as economically regressive, responsible for the energy crisis, and obstructionist by impeding the process by which alternatives (as proposed by energy industries) can be implemented. It is further argued that environmental regulations are inflationary; polluters forced to be responsible for environmental damage pass those costs on to consumers in the form of higher prices.

In 1979 the White House attempted to introduce cost effectiveness as a criterion in the regulatory process by allowing administration economists to comment on the cost-benefit implication of environmental regulations. While cost effectiveness may be a more relevant criterion for assessing internal agency management, such application to federal environmental regulations may be undesirable or at least problematic. In an environmental agency, cost-effective analysis is incompatible with, and not a substitute for, environmental review. Cost-benefit type calculations are arbitrary, dependent upon factors included or omitted in the analysis, the modeling techniques used and the values assigned to variables. However, the pressure is on, and the Environmental Protection Agency (EPA) may become more self-conscious of any inflationary impact of its activities. For example, during 1978 and 1979, the EPA expected to save industry approximately $200 million through relaxed standards on smog and nontoxic wastes from food-processing, glass-manufacturing, and iron-alloy plants.

Part of our dilemma is that we are heavily committed to certain technologies and energy patterns–nuclear power, coal, and synthetic fuels–that are becoming increasingly expensive and damaging to the environment. Over the long term, such a trend is likely to produce major irreversible outcomes. Investment in centralized capital-intensive energy-conversion facilities will become a liability if they are later found to be unsafe, unnecessary, or difficult to close because of the huge amounts of invested capital. If the choice becomes one of investing in capital-intensive, nonrenewable, technologically risky ventures or energy sources with limited environmental impact and unlimited supply, the problem clearly becomes one of finding political means to develop the latter.

## The Political Process

In some situations, there may be irreconcilable differences where trade-offs are difficult to realize. Current administration policy to limit oil for electric power generation has forced some communities to choose between being subjected to low-level radiation from nuclear plants or polluted air from coal-burning plants. Are either the public, or even the experts, sufficiently knowledgeable to evaluate and propose the least-risky future? This no-win dilemma is also a reflection of the governmental process: fragmented policy making, agencies pursuing varying and often conflicting missions, and crisis decision making. The Department of Energy's policy of forcing coal conversion is a direct challenge to the EPA's commitment to prevent the deterioration of air quality. Undoubtedly the EPA may have to make further compromises to conform to the administration's energy program.

Without a comprehensive environmental or energy policy, agencies tend to undermine each others' efforts when their missions conflict. Ultimate policy outcomes rest not on any thoughtful weighing of alternatives but on the relative political strength of an agency (with their attentive constituent and political supporters). These outcomes are more a result of success in bureaucratic politics than of superiority of policy content.

Moreover, agencies themselves have a tendency to increase in size so as to allow internal units to develop autonomy and pursue their specific objectives. Environmental concerns may be ancillary to a public-utility commission or an energy office, yet such agencies are asked to make decisions that have an environmental component. Although some innovation may result from such autonomy, especially on the local level, more often the results are not communicated widely, and agencies find difficulty in pursuing broader goals. That agencies may be working on pieces of the same problem is no guarantee that eventually the pieces will comprise a coordinated policy. Lack of comprehensiveness in policy making is not limited to energy and environmental problems; it is characteristic of a political system that indulges in crisis decision making while postponing problem solving. This indulgence carries with it the penalty of inflicting high costs and restricting options, and that burden falls on a future generation, which eventually must search for solutions.

## Representation of Interests

The crucial question remains: How can political institutions more effectively reconcile energy needs with environmental concerns? Turning to the political and administrative process, a series of concerns come to mind. It must be asked whether all relevant interests have access to decision makers. Who is left out? What is the relative power of contenders? Who in fact are making the decisions, and what is the nature of constituent support? Are the battles being waged as a cover for more-powerful interests? What is the likelihood of minority opinion ever becoming the majority? Are environment and energy conflicts symptomatic of a political and economic system where decision makers are often unaccountable to the electorate?

Many social scientists believe that questions of value are best resolved in the political arena, where the majority rules. Resolutions of energy and environmental issues then rests on the prevailing forces operating in the political arena. But it is only when all interests are represented that one can assume that equitable judgments will be made reflecting public and not private interests. There can be a legitimate role for public participation if the public is informed and administrative bureaucracies allow the public entrance at the early stages of decision making. This does not diminish the need for more-comprehensive analysis as energy and environmental officials consider the interactive effects of their actions.

As agencies become heavily involved in rul making and enforcing regulations, the courts, sometimes reluctantly, have come to play an increasing role. Legislative responsibilities, especially in funding and oversight activities, have given Congress a crucial role with the possibility of influencing executive policy. The potential for action and the interactions of the branches of government in effectuating policy making need greater attention as more governmental units become involved. Other approaches may allow action to be taken at the most appropriate governmental levels. For example, local reforms in zoning regulations and building codes may encourage insulation or energy-saving transportation plans. In other situations, the greater resources and authority of the federal government to deal effectively with environmental or energy problems that transcend political boundaries may be more appropriate.

## Options

Another crucial issue that must be addressed immediately is the overall direction of future policies. Commonly held beliefs about the relationships of energy, environment, and economics must be reexamined. For example, increasing the standard of living does not necessarily mean that the continuation of current energy patterns is inevitable. While estimates of projected electric demand through the 1990s have been lowered because of decreases in current electric demand, we have not yet realized the full potential of conservation and more-judicious use of our energy supplies. There may be more flexibility and greater environmental reward in manipulating the demand for energy rather than the supply. Such efforts have an economic incentive and preserve environmental integrity. However, energy companies have a greater interest in controlling the supply of energy in ways that maintain their preeminence in the energy market. We need to develop a national policy to reconcile energy and environmental goals from a new perspective. Decision makers have the opportunity to provide direction to ensure that the alternatives that we assume will be realized in twenty years will actually occur. Government and utilities have the ability to invest in research and development projects that will keep environmentally benign technologies competitive and widely available, but we have yet to realize the full potential for such alternative investments.

Political mechanisms are available if decision makers consider broader perspectives in selecting criteria for their decisions. Crucial in the shift from fragmented policy making is a more incisive analysis of the effects of discrete actions from a longitudinal perspective. We need to understand how changes in one element of a system will manifest itself in other parts of it. A variety of studies are available that examine only a singular phenomenon restricting the range of discourse. We lack a true marketplace of ideas where informed public officials and citizens can make rational choices directed toward satisfying the broad public interest. Rather we have a cynical, uniformed public signaling to their representatives ambiguous and inconsistent demands. Since governmental policy emanates from crises and has difficulty in dealing with anything but emergencies, government responses are similarly haphazard. The outcomes are policies that reflect the interests of those sectors of the public whose investments are in jeopardy. Such constituencies have greater access to decision-making centers, which then become advocates for their cause.

## Conflict or Resolution?

We are now at a critical juncture. The interactions of energy and environmental issues occur at many levels—political, social, scientific, and economic. The resolution of conflicts between energy and environmental policies and goals represents not only a challenge to our creative potential and innovative technological capabilities but an opportunity to rethink and reshape our political institutions to ensure human survival. No longer can we be content to act only in response to crises; instead we must be prepared to meet crises with developed policies, with long-range perspectives, and with expressed goals. We must be able to utilize the full potential of our political institutions to allow for new directions, which includes critical analyses of our assumptions of energy demand and supply with a fuller and a more realistic appraisal of environmental impact to promote a humane and productive future.

We must ask how political institutions can be made to register public-interest choices where decisions are of a highly technical nature and where decision-making structures are more sensitive to discrete demands. There need not be conflict between energy usage and the environment if a new conception of the future emerges. This is only possible after questions of ethics and values are considered, debated, and resolved whereby people's wants and needs can be made compatible with environmental integrity.

Politics is the art of creating new possibilities for human progress. Therefore we must consider carefully the political framework in which energy and environmental decisions are made. We are confronted with a very uncertain environmental and energy future. We now have the opportunity to shape that future.

# 2 Will Oil Imports Be Available—and at What Price?

*Dankwart A. Rustow*

Petroleum provided most of the drama of world politics in the 1970s. There had been earlier Arab-Israeli wars. There had been previous Soviet-American confrontations in the Middle East and elsewhere. Newly independent governments in the Third World had been overthrown by military coups innumerable times. Iranian masses had been seized by spasms of antiforeign hysteria as far back as the 1890s. But in the 1970s, Qaddafi's seizure of power (1969), the Arab attack on Israel on the morning of Yom Kippur (1973), the Khomeini revolution in Iran (1979), and many lesser events all became inexorably linked to the prospect of embargoes, oil-production cuts, skyrocketing prices, gasoline queues, and hints of invasion of oil fields. The nagging question became, Will Middle Eastern oil be available and at what price?

Let me start by dispelling some legends. One legend asserts that the Organization of Petroleum Exporting Countries (OPEC) is increasing oil prices so as to do us a favor. The ex-shah of Iran used to lecture us on how we wasted oil, "this precious liquid." Some Western observers have echoed the same argument in more sophisticated form. Oil is an exhaustible resource, irreplaceable in human lifetime. Even without OPEC, probably sometime in the twenty-first century, resources would have approached exhaustion; the price would have gone up sharply; and we would have been forced, on short notice, to develop other energy sources to substitute for oil. OPEC is merely hastening that development—raising the price now so as to force us to get busy developing alternative fuels in good time.

The truth is the obvious one: OPEC is increasing prices to get more money. No shortage caused by resource depletion would cause a fivefold jump in price, such as occurred in 1973-1974, or even the tripling that occurred in 1979-1980. The OPEC revolution has meant the greatest short-term redistribution of income in human history, boosting the revenues of member governments from $2 billion when OPEC was founded in 1960 to around $300 billion in 1980. And even though this has contributed to world inflation, there is no question that OPEC has stayed well ahead of any inflation rate.

A second legend holds that the big oil companies have put OPEC up to it. Look at the allegations that keep cropping up of circumvention of price regulations; above all, look at the profits of these companies. For the common citizen caught in a gasoline line in 1979 or facing a doubled fuel oil bill this winter, it does not take much imagination to suspect a grand conspiracy of Exxon, Mobil, Gulf, and other corporate giants.

The truth is far more complex, and it is in three parts. First, until the early 1970s the major oil companies were in full control of oil production in the Middle East and elsewhere and of the global market at large. They used that control throughout the 1960s to drive down the price of oil (from about $2.10 per barrel in 1958 to $1.30 in 1969) and to enlarge their market at the expense of coal. The companies, too, were not acting from altruism. They were working, quite successfully, to increase their sales so as to meet competition from smaller companies, such as Italy's ENI and Armand Hammer's Occidental, and financial pressure from OPEC members while still increasing their own revenues.

Control over oil means deciding how much of it will be produced and how much sold to whom at what price. The companies had arrived in the Middle East when most of the countries were Western colonies, mandates, protectorates, trusteeships, or de facto dependencies. But by around 1970, the vestiges of this imperial control were disappearing. The French left Algeria in 1961; the American air force evacuated Wheelus Field in Libya in 1970 and Dhahran in Saudi Arabia as early as 1962; and the British withdrew from the Persian Gulf in 1971. Middle Eastern governments now felt free to assert their economic self-interest to the limit. And with even the most powerful Western country unwilling to oppose a warlike act such as an embargo, it would have been unrealistic for commercial companies to challenge the sovereign right of OPEC countries to dispose of their subsoil resources. The second part of the truth is that the right to decide how much oil is sold to whom at what price has been solely in the hands of OPEC governments since the early 1970s.

The third element of the truth is that the companies adjusted to this power shift very successfully. They continue as technical experts in charge of production in most OPEC countries and as wholesale buyers of much of the output, and in return they receive a slight discount. They benefit from the arithmetic truth that a 5 percent discount on $10–or even a 2 percent discount on $25–is more than a 20 percent profit on the old price of $2 a barrel. They benefit even more from the geographical and technological fact that it takes six to eight weeks for the oil to get from Middle Eastern wells by tanker, refinery, pipeline, and fuel truck to the neighborhood filling station and from the universal commercial practice of valuing inventory at replacement cost. This means that every time OPEC raises prices, the companies adjust their product prices as quickly as possible, and reap a windfall on the difference. Finally the companies benefit from the economic fact that OPEC is the world's marginal producer and that market prices are set at the margin. Whenever OPEC raises the price, the global cost of all other crude oil goes up apace, at least within the limits permitted by government price regulations.

A third legend has it that we can solve the problem of high prices and shortages by increasing supplies. There has indeed been intense activity in the last decade to add to world oil supplies—from the North Sea, from Alaska, from Mexico. But each of these has added no more than one of OPEC's lesser producers, say Kuwait or Abu Dhabi. In nonrecession years, that is just about the amount that the world has been adding to its energy consumption. In other words, we would have to add another Mexico or Alaska every year—not to break OPEC but to keep supply and demand in their present balance.

Above all, OPEC has been setting the world price, at which Norway, Great Britain, Mexico, and everyone else has been selling their crude oil. And suppose even Mexico turned out to be another Saudi Arabia; that would simply mean that the price-setting power would shift from Riyadh and Tehran toward Mexico City.

A more sophisticated version of the same legend holds that the problem will be solved by developing alternative fuels. This is the rationale behind President Carter's plan in 1979 to apply $88 billion of the proceeds of the proposed windfall profits tax to such projects as synthesizing oil from shale, coal, or tar sands. There is no doubt that part of the long-run answer is to shift from oil toward other forms of energy, including electric power from coal, atomic energy with adequate safety precautions, and solar energy—and not just synthetic fuels.

But let us not be too sure that nonoil alternatives are the way to bring down energy prices. OPEC has been the price setter not only for the global oil market but to some extent for energy in general. From 1972 to 1975, the world price of coal increased two and a half times, that of uranium as much as seven times. The proximate causes were the worst strike by the United Mine Workers since the days of John L. Lewis and a price-fixing conspiracy for uranium of the Canadian government, the Gulf Oil Company, and others. But behind these factors was the great opportunity of the quintupling of oil prices.

With synthetic oil, the coincidences have been even more striking. When U.S. domestic oil sold at about $3 a barrel in the early 1970s, estimates for synthetics ran from $4 to $6. By 1975, when OPEC had raised the oil price to $10, the estimates were $12 to $15. Now that OPEC oil sells at $23 and up, we are told that synthetic crudes can be produced at $25, $30, or $40.

Nonoil substitutes may not be cheaper, and they can be the answer only in the long run. Any really new technology takes a decade or more from research to pilot plant to testing for actual use to development on a commercial scale. The fact that nuclear power after all these decades and tens of billions of dollars supplies no more than 4 percent of our total energy consumption is surely a sobering reflection.

Or take coal, the tried and proven fuel of the industrial age, which was beaten by oil as the noncommunist world's leading energy source as late as 1960. Since then, our energy consumption has tripled, which is to say that production of coal would have to be tripled just to restore it to an equal place with oil. And that implies formidable problems of ecology (strip mining and air pollution) and of transport (railroads, water for slurry pipelines, or radically improved high-voltage transmission technology). Thus, substitutes for oil are our answer only in the very long run.

A fourth legend asserts that our sentimental attachment to Israel is at the root of our Middle Eastern, and hence of our energy, troubles. And it goes on to urge that while guaranteeing Israel's survival, we should abandon all other kinds of sentimental pro-Zionist folly in favor of a policy of friendship with the Arab oil-producing countries. This legend misconstrues the political factors that make up the world oil crisis. If Israel had never come into existence—or alternately, if miraculously a super-peacetreaty were devised that would satisfy everyone—our oil problems would not be substantially different.

The true political factor is that Middle Eastern countries became politically independent of the West just as the West became emotionally dependent on Middle East oil. Other political factors serve either as accidents or as window dressing to enhance the realities of monopoly pricing by a skillful cartel. By political window dressing I mean solemn moralistic declarations such as those accompanying the embargo and production cuts of October 1973. This was not a sacrifice of economic interests for political goals. The production cuts by the Saudis and others were restored by January 1974, long before Israel had evacuated any of the territories acquired in 1967, but after the price of oil had been quadrupled. And this Arab oil weapon was applied not in 1967 when Israel occupied the West Bank, Golan, and Sinai but in 1973 when the entry of the United States into the world market as the major oil importer created an unprecedented boom market. (Between 1970 and September 1973, Saudi government revenue per barrel of oil had doubled, but so had production, a clear signal that the traffic would bear a good deal more by way of prices.)

By political accidents I have in mind events such as the rupture of the Saudi pipeline to the Mediterranean in 1970, which, with the resulting steep rise in tanker rates, gave Qaddafi his first chance for tough bargaining with the companies. And I also have in mind the Khomeini revolution, which, of course, was directed not at the global oil market but at the repressive rule of the shah. But by causing technical and labor difficulties in the oil fields of southern Iran, by first shutting down production and then restoring it at half its former level, the Iranian revolution had a devastating effect on the global oil market by curtailing supplies and increasing prices.

I hope that by now I have dispelled the legends of OPEC as a humanitarian institution to cure us of our profligacy in oil consumption, of the oil companies rather than OPEC as the major problem, of new discoveries in Alaska or Mexico or of oil from shale or coal as the magic solution, and of Israel as the sacrificial lamb that must be placed on the altar of Arab oil interests. The true problem is that there is an oligopolistic structure in the world energy market in which OPEC, non-OPEC oil exporters, the oil companies, the producers of coal and uranium, and the patent holders on shale retorting and coal liquefaction processes all participate. The Middle Eastern oil producers are in the vanguard of the oligopoly, determined to charge what the traffic will bear. Random disturbances endemic to the Middle East—a Qaddafi coup in 1969, a Yom Kippur war in 1973, a Khomeini revolution in 1979—all help to test what the traffic will bear in a far bolder manner than the cautious OPEC oil ministers would normally dare.

It turns out that the traffic will bear a good deal and that there is no early end in sight. Between 1970 and 1973, Saudi Arabia doubled its income per barrel and its production for a fourfold increase in revenues and led the movement to increase the price fivefold. In 1979 Iranian production was cut in half, but the world turned out to be willing to buy that half at twice the price, and OPEC and its individual members vied with each other in raising the price for most of the next year.

But is the situation really that desperate? Can we do nothing except pay, pay, and then pay some more? Must we watch the dollar drop lower and lower, must we see our diplomats being taken hostage and our flag being used to collect trash, all because we are so abjectly dependent on oil from the Middle East? Must we face the prospect of gasoline queues for the rest of your lifetime and mine? Or must we, as the true pessimists tell us, watch the oil wells run dry and then go on to freeze in the dark?

The answer is that there is no need for our dependence on Middle Eastern oil, at least not on the scale on which we have inflicted it on ourselves. There is a way—and indeed a fairly quick and fairly painless way—to cope with our oil problem, and hence with a good portion of our Middle Eastern problems. Small cars are the big answer to America's energy needs. We consume two to two and a half times as much energy per capita as do other industrial nations. If we could keep that excess ratio to one and a half to two times as much, we would need to import no oil whatever.

The greatest disparity is in the transport sector. The Japanese rely on railroads, the Germans on small cars. Of course, we have certain constraints. Our pattern of settlement is far more dispersed than that in Europe or Japan. We are also a more affluent society and therefore more tempted to drive to the store for a pack of cigarettes or to take a fishing trip on the weekend. But if without anyone driving a single mile less, we could manage to attain the European standard of twenty-one miles per gallon (instead of our current performance of fourteen miles per gallon) we could eliminate about a third of our oil imports. If we could improve that standard to thirty or thirty-five miles per gallon, we could do with less than half our present imports.

Our automobile stock turns over every six years or so. Instead of subsidizing Chrysler to rescue it from the effects of poor management and bad market judgment, why not turn the plant over to Volkswagen or Toyota, which already know how to produce economic cars for a profit? Why not cut our oil import bill from the $50 billion to $60 billion, which it was in 1979, and the $75 billion it may be in 1980, and thus strengthen the dollar? Why not take ourselves out of the role as the world's largest oil importer–and thus restore the situation as it was from 1975 through 1978 when world demand for OPEC oil was stagnant or declining–and so was OPEC's price? Small cars are indeed the most effective and the most painless answer during the transition ahead.

The whole might of OPEC<br>
Would be worth not one kopeck,<br>
If only we could toil<br>
With less imported oil.

# 3 The Disincentives for Policy Leadership in Energy and the Environment: The Structure of Voter Opinion

*Nancy Laney* and
*Helen Ingram*

Clearly citizens want their government to act to solve energy problems, yet institutions at both the state and national levels have found it difficult to respond. Legislatures, in particular, have labored long and hard, with few results. Congress has regularly rejected most of what the President has proposed but at the same time has not succeeded in formulating a national energy policy of its own. State legislatures, too, have faltered on energy and environment trade-off questions, often leaving these matters to the executive branch or federal agencies. Overall our representative institutions are not meeting the public's demand for solutions to energy questions.

There are many possible reasons for the painfully slow progress evident in energy and environmental policy. Physical or economic constraints may leave no attractive policy options open to legislators. Intense lobbying efforts on the opposite sides of issues by interests with direct economic stakes and by emotionally charged citizen groups may leave legislators cross-pressured and reticent to act. Certainly it is true that many of the issues are complex and technical and that there is a high degree of uncertainty about the likely consequences of various actions. In an environment with a great many pressures and uncertainties, legislators may look to public opinion for signals about what actions to take. We know that the role of representation becomes most relevant for highly salient, controversial issues such as energy and environmental policy.[1]

In order for constituency opinion to be useful in making policy, however, legislators must be able to characterize it and to frame meaningful responses to it. This chapter examines the structure of public opinion on energy and environmental matters and explores the implications of these attitudes for legislative action. We have chosen to concentrate on state legislatures and voters in the southwestern United States. We will argue that the inconsistent and fragmented attitudes that voters hold about energy and environmental issues make it very difficult for legislators to formulate and support coherent programs. If legislators accurately mirror voters' often inconsistent attitudes, it is exceedingly hard for them to respond to voters' demands for integrated policy.

**The Study**

The data come from a study of voter attitudes on energy and environmental issues in the Four Corners states of Arizona, New Mexico, Colorado, and Utah. This region lends itself to study because it embodies a classic example of potential energy and environmental trade-offs. The area is rich in energy resources at a time when such resources are presumably scarce. Large formations of coal underlie northwest New Mexico, as well as the Navaho and Hopi reservations in northeast Arizona, the Kaiparowits Plateau in Utah, and much of western Colorado. The nation's highest-grade oil-shale deposits are located in northwest Colorado. Uranium is deposited throughout the region in substantial quantities. Demands to develop these resources raise the possibility that the region may become an energy colony for the nation.

Difficulties arise because the area is as rich in natural beauty as in energy resources. Six national parks, twenty-eight national monuments, two national recreation areas, and many state parks and national forests are concentrated here. The region's arid landscape is fragile and slow to recover from human intrusions. Vestiges of roads and mines used over a century ago by pioneers are still clearly visible in many remote stretches of the southwestern deserts. Whether and how the region's energy resources are developed has large implications for the region's delicate physical environment. Figure 3-1 displays the resources and natural areas in these states. Preserving the area's unique environment while at the same time developing resources presents a real challenge.

The appropriate level of government at which to address this challenge is a matter of some debate. Certainly the problems of energy development and environmental protection are important at all levels of government. Energy independence is a national goal, and the economic advantages of energy development are important to many regional, state, and local interests. At the same time, environmental protection has become an issue of increasing concern to state and local governments after virtual domination by the federal government over the last decade. It can be argued that state legislatures are the best available political arena in which to address regional problems of energy development and environmental quality. These state governmental bodies have long-standing constitutional legitimacy and individual legislators have established geographical bases from which to represent constituents. Because of the differences in the scope of electorates, it is reasonable to think that state legislatures have a greater incentive for a careful weighing of the impacts of resource development than does the U.S. Congress, few members of which come from the region. The states are sufficiently small and exclusive to comprehend those citizens who directly benefit and pay the price of development decisions while not including those for whom development impacts are essentially an externality. The Four Corners states, with their vast energy resources, delicate environments, and relatively limited populations, provide a strong rationale for emphasis upon the state legislature in energy and environmental policy.

The data in this chapter come from a larger study of voters and legislators in Arizona, New Mexico, Colorado, and Utah.[2] A sample of fifty voters, stratified according to party registration, was selected from all registered voters in each state legislative district, and questionnaires covering numbers of policy issues were mailed to each voter. Repeated mailings resulted in a response rate of over 70 percent of all deliverable mail. These data were collected during 1975 in Arizona and New Mexico and before the 1976 election in Colorado and Utah. Some questions were asked in only two states. The items from the questionnaires relevant to this study are included in the chapter appendix.

## *Legislative Responsiveness and Constituency Opinion*

Representation is more than simply mirroring the attitudes of constituents. Another overarching concern of representation is to "act in the interests of the represented, in a manner responsive to them."[3] Formulating meaningful issues that can be placed on the governmental agenda, setting priorities among issues, and negotiating compromises necessary to gain acceptance are essential to representation. These go beyond a simple reflection of voter attitudes to the formulation of policies and programs that are responsive to their true interests. Constituency attitudes may serve to facilitate or inhibit this broader kind of responsiveness. Formulating policy is easier when people's preferences are generally compatible within and across issues. Constituent concerns and goals that cut across issues and determine stands on individual questions provide the basis for the development of coherent policy.

Legislators in the process of policy making necessarily need to balance issue positions against one another. If a legislator senses constituency support for a number of linked issues and believes that support may be transitive, then there is considerable flexibility in fashioning policy that is responsive to underlying interests. Inconsistent and fragmented attitudes about issues among voters, however, make it very hard for legislators to frame meaningful responses to problems. Not only is it more difficult for legislators to learn what voters think when issues are perceived as separate and disconnected, but legislative responsiveness is limited to simple issues, and the resulting policies may be incompatible. To understand how voters' opinions help or hinder representatives acting responsively on energy and environmental matters, we must examine the structure of voter opinion to learn something of the distribution, intensity, and consistency of voter attitudes on energy and environmental issues.

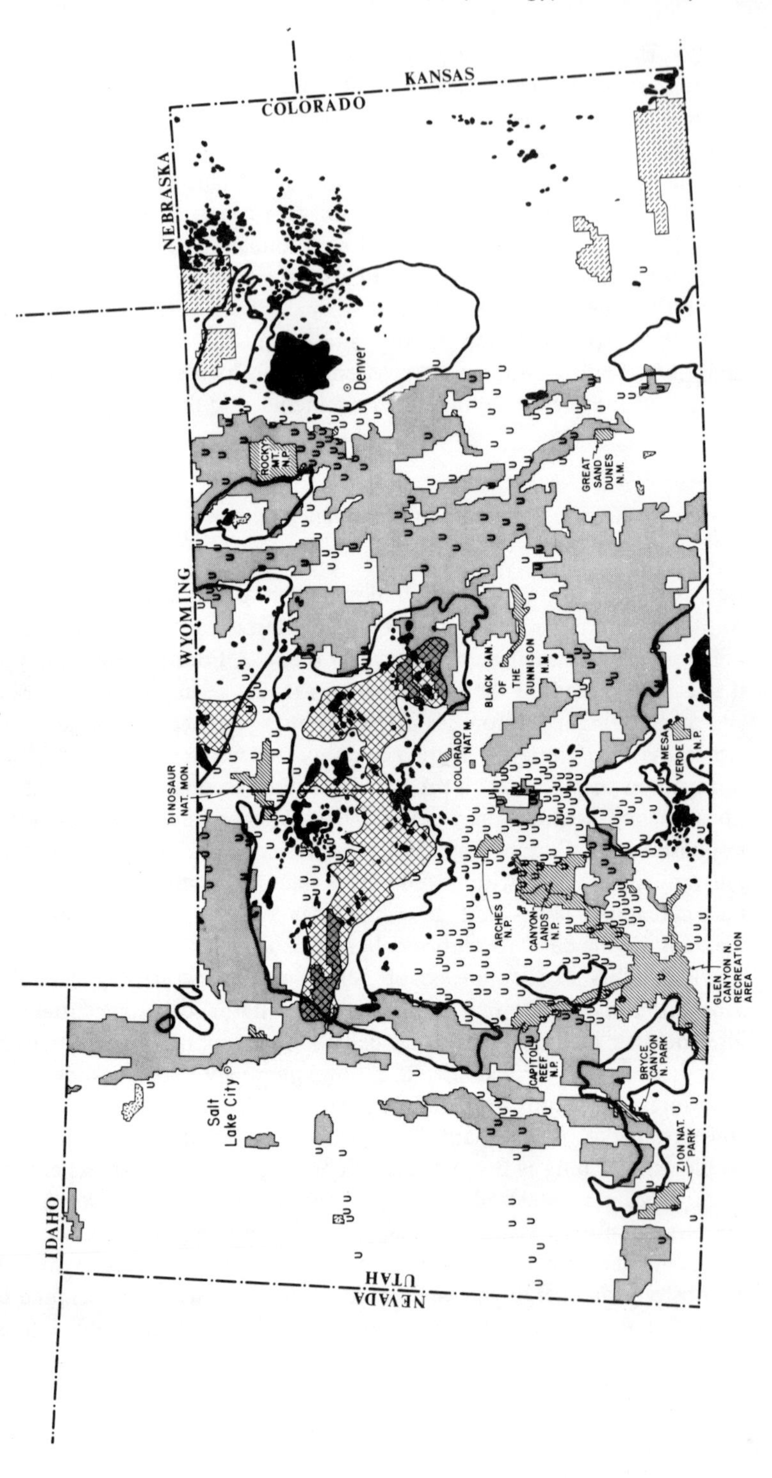
KANSAS
COLORADO
NEBRASKA
WYOMING
Denver
ROCKY MT. N.P.
GREAT SAND DUNES N.M.
BLACK CAN. OF THE GUNNISON N.M.
COLORADO NAT. M.
DINOSAUR NAT. MON.
MESA VERDE N.P.
ARCHES N.P.
CANYON-LANDS N.P.
GLEN CANYON N. RECREATION AREA
CAPITOL REEF N.P.
BRYCE CANYON N. PARK
ZION NAT. PARK
Salt Lake City
IDAHO
UTAH
NEVADA

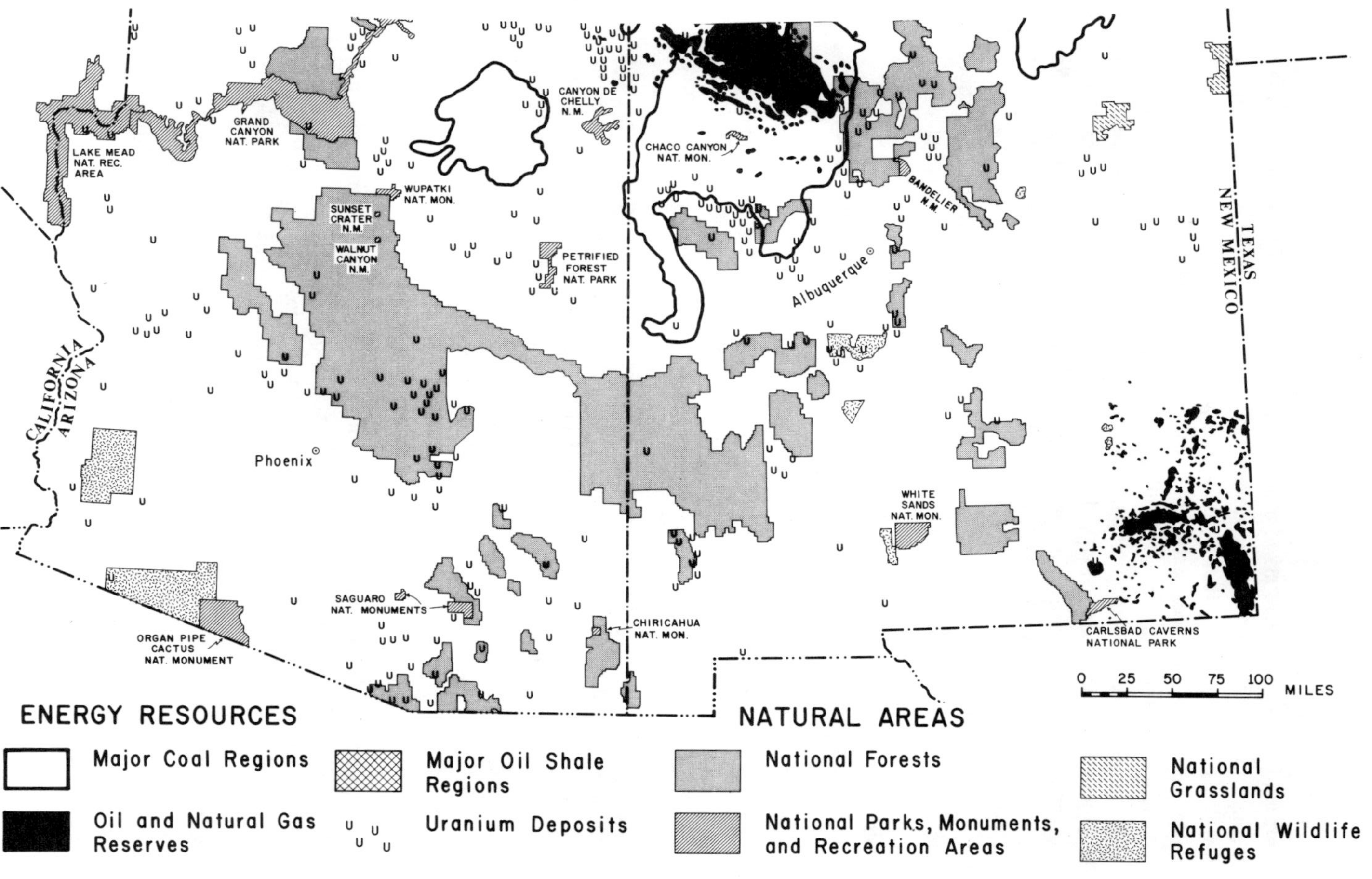


**Figure 3–1.** Energy and Environmental Trade-offs

*Voters' Attitudes on Energy and Environmental Issues*

Voters in our survey were asked a number of questions aimed at tapping their feelings about the seriousness of various environmental problems, their support for the development of their state's energy resources, and their preferences when faced with choices between environmental protection and economic or energy development (see Appendix 3A-1). In asking these questions, we were concerned with determining how important the problems associated with environmental quality and energy development are to the voters in these states and which issues are more important than others. An understanding of the direction and intensity of constituency opinion on these issues is fundamental to representation. Certainly legislators must have a general sense of the wishes of their constituents before they can be responsive to them. At the same time, the attitudes and desires of voters must be constrained by some consistency or logic that allows legislators to generalize across issues and frame meaningful policy responses to problems.

We therefore were also interested in determining the priorities and goals that appear to constrain voters' attitudes about the potentially conflicting values of energy development and environmental quality. In attempting to assess where voters would stand on conflicts involving these values, we were concerned with the consistency of their responses to various questions and their willingness to make explicit choices in trade-off situations. Whether voters recognize that it may often be impossible to achieve both a quality environment and plentiful supplies of energy was a major focus of our analysis. The direction, intensity, and consistency of voters' attitudes on these issues are important to the kinds of signals that legislators receive from their constituents and have important implications for energy and environmental policy in the Four Corners states.

**Support for Environmental Quality**

It is evident that voters are concerned about the seriousness of environmental problems. Table 3-1 shows that in all states but New Mexico, most voters consider air pollution a serious problem. Several cities in the region do have substantial pollution problems. In fact, Denver, Phoenix, and Tucson failed to attain primary air quality standards set by the Environmental Protection Agency. Yet there are still vast areas in the Southwest that traditionally have had very high-quality air, and there is concern that with development this will change. In some areas clean air is now being threatened by both existing and proposed coal-fired plants.[4]

Almost as many people believe that water pollution is just as serious. Except for the increasing levels of salinity in the Colorado River water, water

**Table 3-1**
**Seriousness of Air and Water Pollution**

| | *Arizona* | *New Mexico* | *Colorado* | *Utah* |
|---|---|---|---|---|
| Air pollution | | | | |
| Very serious | 23.5% | 12.3% | 29.4% | 19.1% |
| Serious | 42.0 | 33.8 | 42.7 | 42.0 |
| Not sure | 7.0 | 6.4 | 7.4 | 6.8 |
| Not very serious | 24.4 | 38.9 | 19.0 | 29.2 |
| No problem | 3.1 | 8.6 | 1.5 | 2.9 |
| (*n*) | (1,456) | (2,046) | (1,710) | (1,398) |
| Water pollution | | | | |
| Very serious | 20.6 | 20.1 | 26.7 | 13.8 |
| Serious | 34.7 | 36.5 | 43.1 | 40.4 |
| Not sure | 17.5 | 13.2 | 12.8 | 13.8 |
| Not very serious | 23.2 | 26.0 | 15.6 | 28.5 |
| No problem | 4.0 | 4.2 | 1.9 | 3.5 |
| (*n*) | (1,457) | (2,060) | (1,695) | (1,399) |

pollution is not thought by experts to be an especially troublesome problem in the region. The strong similarity of responses that voters give on air and water pollution, as well as other pollution problems asked in the survey, indicates that voters may be responding negatively to all types of pollution, whether or not they experience them personally.

Voters were somewhat less willing to translate their environmental concerns into a dollars and cents commitment (table 3-2). When they were asked whether pollution control should receive less, the same, or more money from their state legislature in the future, the plurality of voters in all states but Colorado chose the same level of funding. Again except in Colorado, less than 20 percent of the voters thought that pollution control should be one of the three most important areas earmarked for state spending. It is important to

**Table 3-2**
**Support for Spending on Pollution Control**

| | *Arizona* | *New Mexico* | *Colorado* | *Utah* |
|---|---|---|---|---|
| Spend less | 20.2% | 18.0% | 13.6% | 22.1% |
| Spend same | 41.5 | 46.9 | 38.1 | 46.3 |
| Spend more | 38.3 | 35.0 | 48.1 | 31.7 |
| (*n*) | (1,417) | (2,015) | (1,674) | (1,379) |
| Most important area for more spending | 6.3 | 3.9 | 13.4 | 6.5 |
| One of three most important areas | 17.4 | 12.5 | 31.5 | 17.5 |

note that these states in the past have not been particularly concerned with environmental legislation or funding, so a continuation of current spending does not represent a very large commitment.

The greater concern among Colorado voters about the seriousness of pollution as well as their greater willingness to increase funding for its control is related to the large number of environmentalists in that state and the fact that the environment has been a salient policy topic there for a longer time than in the other three states. Colorado also has the most serious air-pollution problem in the Denver metropolitan area.

In contrast to their general desire to spend the same for pollution control in the future, the majority of voters in all four states want more spent on energy research and development in the future (table 3-3). This strong commitment for funding is emphasized by the fact that over one-third of the voters thought that energy research and development was one of the three most important areas for state spending. Energy development appears to be a higher-priority issue than pollution control among voters in the Southwest.

At the same time, voters in these four states are not willing to accept easily the environmental sacrifices associated with large-scale energy development. A majority of voters disagreed with the statement, "We should be willing to accept more air and water pollution in order to insure plentiful supplies of energy." As table 3-4 shows, less than one-quarter of the voters were willing to accept more air and water pollution as the price of energy development.

Voters are even less willing to tolerate environmental costs if the energy that creates the degradation supplies other states. Table 3-5 shows that decided majorities of voters in every state agreed with the statement, "This state should not permit environmental damage in order to produce energy for use in other states." Southwestern voters apparently are not willing to pay the environmental costs associated with seeing their region become an energy exporter.

**Table 3-3**
**Support for Spending on Energy Research and Development**

| | *Arizona* | *New Mexico* | *Colorado* | *Utah* |
|---|---|---|---|---|
| Spend less | 5.0% | 7.7% | 7.0% | 6.2% |
| Spend same | 29.5 | 34.3 | 37.5 | 36.3 |
| Spend more | 65.5 | 58.0 | 55.5 | 57.5 |
| (*n*) | (1,433) | (2,019) | (1,680) | (1,374) |
| Most important area for more spending | 11.6 | 8.1 | 11.4 | 13.2 |
| One of three most important areas | 37.8 | 31.1 | 34.5 | 38.9 |

**Table 3-4**
**More Pollution to Insure Plentiful Supplies of Energy**

| | *Arizona* | *New Mexico* | *Colorado* | *Utah* |
|---|---|---|---|---|
| Strongly agree | 5.4% | 6.0% | 3.8% | 3.8% |
| Agree | 19.4 | 21.8 | 15.8 | 22.2 |
| Not sure | 13.3 | 17.9 | 13.1 | 15.3 |
| Disagree | 43.1 | 39.7 | 45.8 | 44.3 |
| Strongly disagree | 18.8 | 14.5 | 21.4 | 14.5 |
| (*n*) | (1,439) | (2,033) | (1,705) | (1,403) |

## Support for Energy Development

What appears to be a very clear, pro-environment stand among the majority of Four Corner's voters is weakened considerably, however, by their concomitant support of environmentally damaging energy-development strategies. For example, table 3-6 shows that the majority of voters in Colorado and Utah agree that oil-shale development should be encouraged. Large, unexploited oil-shale fields exist in both states, but the extraction technology has not yet been shown to be economically feasible. It is known that large-scale oil-shale development, if undertaken, will be highly water consumptive, with massive land-use requirements and adverse environmental impacts.

Perhaps the adverse environmental impacts associated with oil-shale development are not well recognized or understood by the people in these states. We therefore asked about other energy-development strategies where conflicting values should be more apparent to voters. The first issue concerned the construction of a large thermal-electric-generating station in southern Utah, the Kaiparowits project. This project has generated considerable controversy because of its potential environmental and social impacts and the fact that most of the power generated would be used outside the region. Despite their avowed

**Table 3-5**
**No Damage for Energy Used Outside State**

| | *Arizona* | *New Mexico* | *Colorado* | *Utah* |
|---|---|---|---|---|
| Strongly agree | 31.2% | 34.5% | 27.4% | 22.8% |
| Agree | 37.9 | 36.5 | 36.6 | 37.1 |
| Not sure | 14.4 | 14.4 | 16.7 | 17.7 |
| Disagree | 13.5 | 11.5 | 17.0 | 20.1 |
| Strongly disagree | 3.0 | 3.1 | 2.3 | 2.3 |
| (*n*) | (1,442) | (2,043) | (1,692) | (1,397) |

**Table 3-6**
**Oil-Shale Development Should be Encouraged**

| | *Colorado* | *Utah* |
|---|---|---|
| Strongly agree | 13.6% | 22.4% |
| Agree | 41.4 | 54.9 |
| Not sure | 29.9 | 11.5 |
| Disagree | 10.7 | 3.8 |
| Strongly disagree | 4.4 | 1.3 |
| (*n*) | (1,695) | (1,399) |

Note: The question was not asked in Arizona and New Mexico.

unwillingness to accept more pollution for energy or to suffer damage for energy to be exported, the majority of Utah voters favor this project.

Although the Kaiparowits was the most well-publicized energy issue in the region at the time of our survey, the data presented in table 3-7 show that most Colorado voters were not sure about whether it should be constructed, and even a quarter of the Utah voters took a neutral position on the project. The Colorado voters may consider the project to be Utah's problem and therefore of little consequence to them. If that is in fact the case, it belies a regional perspective about energy development and environmental protection among voters in the Southwest. With the limited populations and congressional representation in these states, a regional focus on these matters has long been considered essential. The fact that the most environmentally aware voters in our sample, those in Colorado, generally did not have any opinion about a large coal-fired plant in neighboring Utah indicates instead an attitude of each state for itself.

A potential hazard related to the development of large energy projects in sparsely populated areas of the region are boomtowns.[5] These "overnight

**Table 3-7**
**Kaiparowits Power Plant Should be Constructed**

| | *Colorado* | *Utah* |
|---|---|---|
| Agree | 19.7% | 50.8% |
| Not sure | 69.4 | 26.0 |
| Disagree | 10.9 | 23.2 |
| (*n*) | (1,637) | (1,400) |

Note: The question was not asked in Arizona and New Mexico.

cities," which spring up with the availability of construction and other project-related jobs, are plagued by shortages of services and make heavy demands on taxpayers. Because of their rapid growth, lack of planning, and heavy dependence upon a single industry, these towns tend to be a boom-and-bust phenomenon in the areas where large-scale energy development occurs. Table 3-8 shows that voters in Colorado and Utah are not particularly concerned with the problem of boomtowns, however. Perhaps the adverse social and economic consequences of such communities are mainly in the future and will be felt only if and when large-scale energy development occurs. The proportion of voters who say they are not sure about boomtowns indicates that geographically distant problems, remote from most voters in these states, may be less relevant to evaluate.

Few nuclear-power plants currently exist in the Four Corner states, but the issue has received considerable national debate. A referendum requiring stringent safeguards for nuclear-power plants was defeated in Colorado and Arizona in 1976 by a division close to that found in our survey conducted before the elections.[6] Table 3-9 shows that although a number of the voters are not sure, most agree that the possible benefits from a nuclear-power plant far outweigh the possible hazards. (This survey was conducted before the accident at Three Mile Island.)

The Kaiparowits and nuclear-energy responses seem to indicate that the costs of some proposals to produce energy are acceptable. At the same time, Four Corners state voters and senators are concerned with the adverse effects of energy production. The strip mining of western coal is essential for its energy development yet has severe land-use, water-use, and environmental implications. A national strip-mine reclamation bill, which would require that land be returned approximately to its original condition after stripping, a costly process, had been debated in the West for several years prior to our survey. Strip mining is not yet widely employed in the region, and the copper and coal

**Table 3-8**
**Seriousness of Boomtowns**

| | *Colorado* | *Utah* |
|---|---|---|
| Very serious | 6.6% | 5.2% |
| Serious | 17.1 | 15.2 |
| Not sure | 36.4 | 28.4 |
| Not very serious | 26.1 | 32.7 |
| No problem | 13.8 | 18.5 |
| (*n*) | (1,674) | (1,391) |

Note: The question was not asked in Arizona and New Mexico.

**Table 3-9**
**Possible Benefits of Nuclear Power Plants Outweigh Possible Hazards**

| | *Arizona* | *New Mexico* | *Colorado* | *Utah* |
|---|---|---|---|---|
| Strongly agree | 25.6% | 15.1% | 18.6% | 17.1% |
| Agree | 29.8 | 34.6 | 30.7 | 33.1 |
| Not sure | 31.5 | 36.6 | 35.3 | 33.6 |
| Disagree | 8.7 | 8.2 | 9.8 | 13.5 |
| Strongly disagree | 4.4 | 5.4 | 5.5 | 2.7 |
| (*n*) | (1,438) | (2,038) | (1,675) | (1,398) |

strip mines that do exist tend to be in more remote areas away from population centers. The adverse environmental impacts associated with strip mining are widely known, however, and it is unlikely that respondents to our survey were unfamiliar with this aspect of energy development. Table 3-10 shows little agreement on the issue. Many voters are not sure. And although substantial percentages of voters see strip mining as not serious or no problem, as many or more consider it a serious or very serious problem. Only in Utah do those not concerned with the problem outnumber those who are concerned.

### Signals to Legislators

People in the Four Corners states generally believe that environmental problems in their states are serious and that environmental damage is too high a price to pay for energy production, especially if the energy will supply other states. Yet the majority of the same people support energy-development strategies that have severe environmental implications and that involve producing energy for export outside the region. Oil-shale development, the

**Table 3-10**
**Seriousness of Strip Mining**

| | *Arizona* | *New Mexico* | *Colorado* | *Utah* |
|---|---|---|---|---|
| Very serious | 16.0% | 13.4% | 16.3% | 7.3% |
| Serious | 25.1 | 21.2 | 25.8 | 15.8 |
| Not sure | 31.6 | 36.8 | 31.9 | 38.0 |
| Not very serious | 17.2 | 19.2 | 20.1 | 29.0 |
| No problem | 10.1 | 9.5 | 6.0 | 9.9 |
| (*n*) | (1,447) | (2,040) | (1,672) | (1,393) |

Kaiparowits coal-fired electric-generating station, and nuclear-power development all receive the go-ahead from Four Corners' voters. Even the problems of boomtowns and strip mining do not appear to be of particular concern to these voters. Apparently these people have not linked the environmental problems associated with energy development with their desire for energy production.

The signals to state legislators on energy and environmental matters represented by these responses are mixed. Legislators attempting to resolve difficult policy dilemmas involving environmental costs and benefits from increased energy production will find little to guide them in these constituency attitudes.

### *Willingness to Make Trade-offs with a Scarce Resource*

It may be, however, that voters are willing to make tough choices involving trade-offs when particular, highly valued resources are at stake. In the arid Southwest, water is certainly such a resource. An inscription on the wall of the Colorado capitol states, "Here is the land where life is written in the water." Water is indeed a scarce resource in the region and a key factor necessary to energy development, other kinds of economic development, and a continuation of rapid population growth.

Historically between 80 and 90 percent of surface and groundwater used in the Four Corners states has gone to irrigated agriculture. In some areas, however, cities threaten agriculture's long-term water supply. Water demands for energy development compound urban pressures upon irrigated agriculture. Every possibility for energy development uses water. It has been estimated that a one-million-barrel-per-day oil-shale development in eastern Colorado or western Utah would consume approximately 150,000 acre-feet of water per year. A coal-gasification plant in New Mexico or Arizona processing 24 million tons of coal per year to meet the energy needs of a million people would use about 300,000 acre-feet of water per year. A 10,000 megawatt coal-fired thermal electric power plant in the Four Corners region would consume about 230,000 acre-feet per year.[7] Proposed coal slurry pipelines, using water to transport finely crushed coal to power plants in other states, would also require substantial amounts of water.

Currently energy production consumes less than 3 percent of available water supply in the Four Corners states, but its share is likely to increase dramatically as development proceeds. Because water is an essential and relatively inexpensive input in most energy processes, the energy industry is likely to be aggressive in securing whatever water supplies it needs for development. Weatherford and Jacoby argue that if one includes both projected and planned energy development in the Colorado River basin, the total projected demand for water may well exceed surface supply in a decade.[8] Further, water consumption for energy will mean greater salinity in the already saline Colorado River and a decrease in the usefulness of its waters for some purposes. Table 3-11 shows a recognition among the majority of voters that the shortage of water is a serious problem in the region.

Given the voters' concern about water shortages in the region, it is important to know whether voters are willing to make choices among the various water uses, including energy production, in allocating this scarce resource. In our questionnaire we asked, "Water use is also an issue of importance in our area. Indeed, the Southwest may eventually have to set priorities among various water users. In your opinion, should each of the following users get *more*, the *same*, or *less* water in the future?" Table 3-12 displays voter opinion concerning the future water allocation for energy. Over one-third of the voters believed that energy should receive more water in the future, and somewhat less than one-half of the voters thought that energy should receive the same amount. Fewer than 5 percent of the voters wanted energy to receive less water in the future.

Taken alone, these responses appear to represent a real commitment to energy, especially considering the voters' previously expressed concern about water shortages. However, the voters were even more in favor of increasing water allocations to irrigated agriculture, which currently uses the most water in the region. Table 3-13 reports the percentage of voters responding that each user should get more or the same amount of water in the future. Clearly despite their concern about the scarcity of water, voters are reluctant to make choices among users. There are some differences in the priority given to certain users in that water-based recreation and Indians fare less well with voters than do other users. The overriding message, however, is that all users should get more or the same amount of water in the future and that no one should have to sacrifice.

### *Signals to Legislators about Water*

It is unlikely that the dwindling supplies of water available in the Southwest can be stretched to serve increased energy production without forcing other

**Table 3-11**
**Seriousness of Water Shortages**

| | *Arizona* | *New Mexico* | *Colorado* | *Utah* |
|---|---|---|---|---|
| Very serious | 27.1% | 24.7% | 24.9% | 13.6% |
| Serious | 40.8 | 36.9 | 41.7 | 36.4 |
| Not sure | 17.2 | 15.8 | 15.7 | 17.2 |
| Not very serious | 11.8 | 16.2 | 15.9 | 26.6 |
| No problem | 3.1 | 6.4 | 1.7 | 6.2 |
| (*n*) | (1,451) | (2,089) | (1,703) | (1,409) |

**Table 3-12**
**Future Water Allocations for Energy Production**

| | *Arizona* | *New Mexico* | *Colorado* | *Utah* |
|---|---|---|---|---|
| More | 42.3% | 33.0% | 35.5% | 43.1% |
| Same | 45.2 | 50.1 | 50.5 | 43.2 |
| Less | 6.3 | 5.1 | 4.5 | 4.9 |
| No opinion | 6.2 | 11.8 | 9.5 | 8.9 |
| (*n*) | (1,396) | (1,992) | (1,646) | (1,348) |

users to reduce their water use. We had undertaken this analysis of voters' attitudes about water and water reallocation in the hope that voters would be willing to make trade-offs involving a specific resource that they would not make with more general values such as environmental quality. That did not prove to be the case. Instead voters displayed a clear unwillingness to make choices among water users. Despite their concerns about shortages of water, the voters in these states generally were not willing to reallocate water away from any current users to provide for other developing uses of water. The inherent conflict between water scarcity and the same or more water for all users in the future does not appear to constrain voters' attitudes.

Legislators who are attempting to respond to the problem of water scarcity though negotiation and compromise will find little support from the public. The voters' inconsistent and particularistic attitudes about water do not encourage the development of coherent water policy where some users will be forced to cut back. It is difficult, in fact, to fathom how legislators are to be responsive to the conflicting attitudes of voters about water. Apparently the voters' inconsistency in responding to energy and environmental

**Table 3-13**
**Future Water Allocations: Percentage of Voters Responding "More" or the "Same" Amount of Water in the Future**

| | *Arizona* | *New Mexico* | *Colorado* | *Utah* |
|---|---|---|---|---|
| Electrical energy production | 87.6 | 83.1 | 86.0 | 86.2 |
| Irrigated agriculture | 91.6 | 94.6 | 92.7 | 92.9 |
| Industry and manufacturing | 84.8 | 79.3 | 76.8 | 84.2 |
| Municipal and residential | 90.6 | 88.0 | 86.4 | 92.3 |
| Water-based recreation | 59.9 | 60.2 | 60.2 | 62.6 |
| American Indians | 80.9 | 75.3 | 66.2 | 72.2 |

questions is not an isolated phenomenon but instead is indicative of a fragmented and particularistic way of looking at environment, development, and resource issues.

## Conclusions and Implications for Policy

We found that the people in the Four Corners states are quite concerned about a number of environmental problems and, moreover, say they are not willing to see the environment damaged in order to increase energy production. Voters are especially opposed to damage that may result from the production of energy to be used in other states. One-third of the voters are willing to back up their concerns about environmental problems with an acceptance of higher levels of government spending for pollution control.

Higher on the governmental agenda for increased spending among these respondents is energy research and development, however. There is also substantial support among the voters for various energy-development strategies under discussion in the region. Voters in Colorado and Utah are generally very supportive of oil-shale development, voters in Utah support Kaiparowits, a large coal-fired electric-generating plant in a beautiful and unspoiled area of their state, and voters in all four states generally support nuclear-energy development. About equal numbers of voters see strip mining as not serious as see it as a serious problem, and very few voters in Colorado and Utah are concerned about boomtowns.

Another major concern of our analysis was to assess voters' opinions on questions about conflicts between environmental protection and energy development. The opinions expressed by voters, when viewed separately by issue, appear to establish clearly their preferences for either environmental quality or for energy development, depending upon the particular question. The difficulties arise as one looks across issues and attempts to determine overall goals or preferences expressed by the voters. Their responses indicate that they are not consistently pro-development or pro-environment. The majority of voters in these states hold a number of contrary opinions on energy and environment trade-off issues indicative of fragmentation and disassociation of many issues from seemingly related ones.

We found that the voters' reticence to make trade-offs among competing values extended even into the area of water policy. The majority of voters in all four states feel that the shortage of water is a serious problem, but when asked to choose among competing uses of water they are unwilling to reduce allocations to any user. The obvious contradiction between the recognition of water scarcity and the desire that all users should get the same or more water in the future does not appear to constrain voters' attitudes about water.

The picture that emerges of public opinion in the Southwest on energy and environmental issues can best be described as a collage. Disparate and even conflicting attitudes about environmental protection and energy development exist side by side in the minds of the voters, and they give little thought to overall consistency. Issues are seen in fragmented and particularistic ways that make linkages difficult and generalizations nearly impossible. There seems to be not so much a lack of information about issues on the part of voters as a failure to look across issues and resolve conflicts by making difficult trade-off decisions.

The task of state legislators in formulating energy and environmental policy is complicated by this collage of voter opinion. Inconsistent and particularistic attitudes among a legislator's constituents mean that he or she will be getting very mixed signals about how to act. The ability to construe voters' wishes and act in a manner responsive to them requires that those attitudes be generally consistent and logically related to available policy options. Yet this study of Southwest voters' attitudes on environmental and energy matters has revealed that they are fragmented, often contradictory, and in some cases beyond the realm of what it is possible for any legislature to deliver.

These findings do not portend well for our representative institutions in meeting the public's demand for solutions to energy questions. Ironically the attitudes held by the public about the issues surrounding energy development seem to be a major obstacle to positive action by their representatives. Support for tough, comprehensive policy choices designed to resolve conflicts between energy development and environmental quality does not appear to exist among voters in the Southwest. Without that support, legislators have little incentive to risk the ire of special interests and perhaps even the general public with comprehensive and conflictual policy choices.

It is much more likely that energy and environment trade-off questions will continue to be fought out on a piecemeal, issue-by-issue basis as they have been in the past. Since support for one issue cannot automatically be translated into support for seemingly related issues, the task of coalition building and negotiation on energy and environmental issues is continuous. Some of these issues will undoubtedly evoke intense support or opposition among particular groups or interests that stand to gain or lose depending upon the final policy outcome. Such political situations are likely to result in intensive, undisciplined lobbying efforts, with large numbers of group representatives making their cases separately before legislators. On less-contentious issues, interest groups with fairly specific clienteles will lobby separately or in loose alliances on individual measures or specific provisions of measures. In either case, the emphasis will be on separate issues and piecemeal solutions.

In acting on these individual energy and environment issues, state legislators in the Four Corners states can be expected to choose symbolic rather than substantive responses to problems whenever possible.[9] These could include such symbolic actions as public statements, formal resolutions, and preambles to legislation that express regard for environmental quality or commitment to energy self-sufficiency. Such actions are far less risky than substantive decisions that generate opposition and for which public support may not exist. Although it is unlikely to produce coherent policy, the conduct of politics in symbolic terms may perform the function of elucidating values and defining moral commitments.

When substantive action is unavoidable, legislators can be expected to take actions that minimize their costs. The costs of energy and environment decisions are likely to be high. Gathering the necessary information about these issues, which are often complex and technical, will be costly. Building coalitions in favor of action will also be costly. Coming to agreements across issues that the public sees in discrete and particularistic terms will make bargaining very difficult on energy and environmental matters. Salisbury and Heinz have postulated that where the cost of allocative decisions (where legislators themselves distribute benefits) are high, legislatures are likely to make structural decisions (where decision-making responsibilities are delegated).[10] State legislatures may leave decision-making responsibilities to the federal government or give vague and general mandates to executive agencies that administrators and eventually courts must interpret. Very little substantive environmental policy has come from these state legislatures in the past, and without support for change from the public there is little constituency incentive to do so in the future.

## Notes

1. Richard W. Boyd, "Popular Control for Public Policy: A Normal Vote Analysis of the 1969 Election," *American Political Science Review* 66 (June 1972): 429-449.

2. Analysis of the complete data set may be found in, Helen Ingram, Nancy K. Laney, and John R. McCain, *A Policy Approach to Political Representation: Lessons from the Four Corners States* (Washington, D.C.: Johns Hopkins University Press for Resources for the Future, 1980).

3. Hanna Pitken, *The Concept of Representation* (Berkeley and Los Angeles: University of California Press, 1967), p. 209.

4. E.G. Walther, W.C. Malm, and R.A. Cudney, *The Excellent but Deteriorating Air Quality in the Lake Powell Region*, Lake Powell Research Project Bulletin, No. 52 (Los Angeles: University of California at Los Angeles, Institute of Geophysics and Planetary Physics, 1977).

5. B.C. Ives, W.D. Schulze, and D.S. Brookshire, *Boomtown Impacts of Energy Development in the Lake Powell Project*, Lake Powell Research Project Bulletin, No. 28 (Los Angeles: University of California at Los Angeles, Institute of Geophysics and Planetary Physics, 1976).

6. The pro-nuclear position in Colorado received 70.9 percent of the

vote, while our survey data indicated that 76.3 percent of the respondents (recoded to omit "not sures") thought that the possible benefits outweighed the hazards. In Arizona, the corresponding percentages were 70.1 and 80.8.

7. Dean E. Mann, *Implications of Changed Water Demand and Utilization*, Lake Powell Research Project Bulletin, No. 27 (Los Angeles: University of California at Los Angeles, Institute of Geophysics and Planetary Physics, 1977).

8. Gary D. Weatherford and Gordon C. Jacoby, "Impact of Energy Development on the Law of the Colorado River," *Natural Resources Journal*, 15 (January 1975): 171-214.

9. Murray Edelman, *The Symbolic Uses of Politics* (Urbana: University of Illinois Press, 1964).

10. Robert H. Salisbury and John Heinz," A Theory of Policy Analysis and Some Preliminary Applications," in *Policy Analysis in Political Science*, ed. Ira Sharkansky, p. 47 (Chicago: Markham, 1970).

# Appendix 3A

Q-25 One of the most important things a state legislator decides is where to spend the taxpayers' money. For each of the issues listed below, please indicate whether you would favor your state legislator spending *less* money on these issues, the *same* amount, or *more* money than is now being spent? (Circle your answer)

| | *Spend* | *Spend the* | *Spend* |
|---|---|---|---|
| 3. Pollution control | Less | Same | More |
| 7. Energy research and development | Less | Same | More |

Q-30 One particular set of issues that has received a great deal of attention in the past few years concerns the environment. The following items are thought by many to be environmental problems in this state. Others do not agree. For each problem do you feel it is very serious, serious, not very serious, or no problem at all in this state?

| | | | | | |
|---|---|---|---|---|---|
| 1. Air pollution | Very Serious | Serious | Not Sure | Not Very Serious | No Problem |
| 2. Water pollution | Very Serious | Serious | Not Sure | Not Very Serious | No Problem |
| 3. Traffic congestion | Very Serious | Serious | Not Sure | Not Very Serious | No Problem |
| 4. Littering | Very Serious | Serious | Not Sure | Not Very Serious | No Problem |
| 5. Population growth | Very Serious | Serious | Not Sure | Not Very Serious | No Problem |
| 6. Strip mining | Very Serious | Serious | Not Sure | Not Very Serious | No Problem |
| 7. Housing developments in rural or undeveloped areas | Very Serious | Serious | Not Sure | Not Very Serious | No Problem |
| 8. Water shortages | Very Serious | Serious | Not Sure | Not Very Serious | No Problem |
| 9. Soil erosion | Very Serious | Serious | Not Sure | Not Very Serious | No Problem |

| | | | | | |
|---|---|---|---|---|---|
| 10. Boomtowns | Very Serious | Serious | Not Sure | Not Very Serious | No Problem |

Q-31 People have different ideas about how our natural resources should be used and conserved. Would you please indicate the extent to which you agree or disagree with each of the following statements.

| | | | | | |
|---|---|---|---|---|---|
| 4. We should be willing to accept more air and water pollution in order to insure plentiful supplies of energy. | Strongly Agree | Agree | Not Sure | Disagree | Strongly Disagree |
| 5. Oil shale development should be encouraged. | Strongly Agree | Agree | Not Sure | Disagree | Strongly Disagree |
| 6. The possible benefits from a nuclear powered electrical plant far outweigh the possible hazards. | Strongly Agree | Agree | Not Sure | Disagree | Strongly Disagree |
| 7. This state should not permit environmental damage in order to produce energy for use in other states. | Strongly Agree | Agree | Not Sure | Disagree | Strongly Disagree |
| 10. The Kaiparowits power project in southern Utah should be constructed. | Strongly Agree | Agree | Not Sure | Disagree | Strongly Disagree |

Q-33 Water use is also an issue of importance in our area. Indeed, the Southwest may eventually have to set priorities among various water users. In your opinion, should each of the following water users get *more*, the *same*, or *less* water in the future? (Circle your answer)

| | | | | |
|---|---|---|---|---|
| Electrical energy production | More | Same | Less | No Opinion |
| Irrigated agriculture | More | Same | Less | No Opinion |
| Water-based recreation (fishing, swimming, boating, etc.) | More | Same | Less | No Opinion |
| Industry and manufacturing | More | Same | Less | No Opinion |
| Municipal and residential uses | More | Same | Less | No Opinion |
| American Indians | More | Same | Less | No Opinion |

# 4 Congress and National Energy Policy: Assessing the Policymaking Process

*Michael E. Kraft*

On July 15, 1979, following several months of gasoline scarcity and sharply increasing prices, President Jimmy Carter took his case for a national energy policy to the American public. It was not his first appeal for public support, nor would it be his last. Speaking at length about the "crisis of confidence" in American values, the president recommended new actions intended to reduce American dependence on OPEC oil and to achieve "energy security." His proposals focused on the development of a synthetic-fuels industry and creation of an energy mobilization board to "cut through the red tape, the delay and the endless roadblocks to completing key energy projects."[1]

Reactions to the speech indicated that public and congressional approval was at best uncertain. The *New York Times* called the president's proposals "timid"; environmentalists termed the "synfuels" program dangerous; antinuclear activists expressed no great liking for the energy mobilization board; and Senators Edward Kennedy and John Durkin introduced a counterproposal to the Carter package that emphasized conservation and solar-energy sources.[2]

These events serve as a useful reminder that the development of a national energy policy will continue to dominate the American political agenda for some time and will necessarily depend upon the resolution of complex and intensely controversial issues of economics, technology, and environmental protection. Should the federal government institute a large-scale national program of energy conservation? Should it encourage substantially increased domestic energy production? Should emphasis be placed on renewable sources like solar energy or on more conventional sources like nuclear energy and fossil fuels? What particular programs should be created to achieve those goals? What costs and risks are associated with each alternative, and who should bear the costs and assume the risks? Technical information about benefits, costs, and risks can help to inform such choices, but the decisions about what mix of energy programs is acceptable in the short run and what type of energy future is most desirable in the long run are political, not technical, decisions. Unfortunately events of the past few years have demonstrated how difficult it is to make those decisions when there is no public consensus on energy policy goals and when the political process frustrates the building of such consensus.

During 1977 and 1978, the U.S. Congress deliberated on President Carter's national energy plan. It was the most ambitious and far-reaching package of energy legislation ever sent to Capitol Hill, a model of the comprehensive and integrated approach so often favored by policy analysts and planners. Yet Congress rejected the key elements of the plan; the bill that finally emerged from the legislative process fell considerably short of a national energy policy. What were the major decisions made on the Hill? What were the reasons for this congressional reaction to the Carter plan? What do the decisions during this period tell us about probable congressional reaction to President Carter's policy proposals of July 1979 and to similar proposals in the future?

A survey of the energy-policy process of 1977-1978 allows some modest speculation about congressional policy making in the near future. I am also concerned here with the capacity of existing political institutions and political processes to resolve in some reasonably democratic fashion the nation's exceedingly complex energy problems. What does the outcome in this case imply about the nation's capacity to deal with continuing energy issues?

## The Policy Process

The policy-making process is usually described in terms of a series of stages in which critical activities and decisions take place: setting the societal and political agendas, formulating policy proposals, legitimizing and adopting public policy, implementing the policy, evaluating program operations and accomplishments, and recommending policy change.[3] For present purposes, the first three stages are the most significant because the nation has yet to adopt an energy policy.

### *Agenda Setting*

Energy issues rose to their current prominence on both the societal and governmental agendas with the 1973-1974 Arab oil embargo and the subsequent quadrupling of prices of OPEC oil. Prior to 1973, energy was not an issue in American politics. There was no Department of Energy, no energy committee in Congress, and no overwhelming concern among policy makers or the public. Energy abundance and cheap prices were assumed likely to continue indefinitely. There were, of course, a variety of discrete federal energy policies, but these largely involved price regulation and allocation of public resources within the various segments of the energy industry (coal, oil, natural gas, and electric power production). On the whole, policy making was fragmented, disjointed, and incremental.[4]

Following the 1973-1974 embargo, there were a number of presidential and congressional actions on energy. As the implications of rising national demand for energy and the costly and uncertain supply of nonrenewable fossil fuels (particularly oil) became clearer, Presidents Nixon and Ford proposed a series of modest changes in energy policy. In November 1973, President Nixon announced Project Independence; the goal was self-sufficiency in energy in the United States by 1980. Widely criticized as impossible to achieve, the goal was later redefined as independence from insecure foreign sources of oil by 1985. The measures proposed to reach this goal included a return to year-round daylight savings time, a reduction of the national speed limit to fifty miles per hour, additional funding for energy research and development, relaxation of environmental regulations (especially clean-air standards for coal conversion), deregulation of natural-gas prices, and authority to implement gasoline rationing and emergency conservation measures.[5]

In 1975 President Ford proposed a thirteen-part energy independence bill. It included measures to raise import fees on imported oil by three dollars per barrel, to lift federal controls that held down the price of domestic oil, to require energy-efficiency labeling on all major appliances and automobiles, to authorize development of and production of oil from the several federally owned oil reserves, to deregulate the price of new natural gas, and to delay deadlines for compliance with clean-air regulations.[6] The major goal of the Ford proposals was to cut energy consumption and to stimulate domestic energy production by relying on higher fuel prices and the normal operation of the market economy to provide the necessary incentives for consumers and producers to change their behavior.

In addition to these major policy proposals, the 1973-1976 period saw the development of a number of private analyses and recommendations, some going considerably beyond the Nixon and Ford actions. For example, in 1974 a Ford Foundation study called for a national commitment to a soft-energy future and adoption of major conservation policies to achieve that goal. It was well received by those partial to the conclusions and severely criticized by others for its questionable economic assumptions, as well as for its support of a strong and coordinated federal energy policy.[7]

A brief summary statement on all of these proposals and the various presidential and congressional actions cannot begin to do justice to the complexity of the issues and the efforts of all involved. However, it seems fair to suggest that the net effect was very little progress in the direction of a national policy that would, in fact, reduce dependence on foreign oil by dramatically altering consumption and production patterns. One review of the Nixon and Ford administration programs characterized them as minor and chiefly symbolic in nature, a "nonpolicy" for energy.[8] But Congress must share in the blame. *Congressional Quarterly* was not alone in suggesting that Congress demonstrated a remarkable inability to overcome its internal fragmentation and the divisive regional and political differences on the issues.[9] Its performance during the 1973-1976 period was repeated to a large extent during 1977-1978.

Following this first period of increased concern and activity, gasoline supplies returned to normal and winter heating oil became easily available once again, although both gasoline and oil were substantially more expensive. Not surprisingly, public concern declined fairly rapidly, and energy consumption continued to rise by 5 percent per year. This change is of special importance because national policy making depends very heavily upon a high level of public concern, which helps provide the political incentives so critical to the formulation of innovative public policy and to the building of political support for policy adoption. In the absence of a national energy policy and with continued increases in public consumption of energy, the United States imported over 50 percent of its oil by 1977. This continued dependence on OPEC nations for half of our oil supply, and at an ever-increasing and damaging cost, set the stage for President Carter's proposal of the national energy plan of 1977.

On April 18, 1977, the president addressed the nation about "a problem unprecedented in our history," which he termed "the greatest challenge our country will face during our lifetimes." He went on to say that the decision on energy policy "will test the character of the American people and the ability of the President and Congress to govern this nation. This difficult effort will be the 'moral equivalent of war'—except that we will be uniting our efforts to build and not to destroy."[10] On April 20 Carter delivered a televised address to a joint session of Congress, outlining his legislative program for energy. A televised news conference followed on April 22. Finally on April 29, the national energy plan itself, a comprehensive and complex package of regulatory and tax legislation emphasizing incentives for conservation of energy and a shift from oil and natural gas to coal, was sent to Congress.

*Formulation*

There is no single approach used in formulating public policy and no consensus on how the process should work, but there are a number of common standards, substantive and political, that are often used to assess policy proposals and the process used to develop them. For example, on a substantive or technical basis, there is usually concern with the quality of data generated or used, the validity of assumptions made in reviewing policy options (for example, on technical feasibility), and the accuracy of calculations (for example, on economic costs and benefits). On a political basis, the process of policy formulation is often evaluated in terms of the scope of the analysis of the problem, the value judgments made on policy goals, the quality of representation of relevant interests, and the validity of assumptions about political feasibility. Formulation of the national energy plan in early 1977 can be criticized in terms of many of these standards.

The plan was assembled by a small task force of fifteen economists, lawyers, and public administrators under the direction of Secretary of Energy

James Schlesinger.[11] With a support staff of twenty-five secretaries and clerks, the task force worked under intense pressure to meet a Carter-imposed deadline of April 20. Their backgrounds were largely governmental service and university teaching, and in contrast to the Nixon-Ford energy planners, they emphasized the role of government rather than of business in meeting the nation's energy shortage. They also looked more to conservation than to increased domestic supplies for the solution. Apparently the president requested that the task force prepare the best possible analysis and policy recommendations regardless of political ramifications. Such advice is consistent with President Carter's style of decision making, which tends to emphasize comprehensive internal analysis of a problem, selection of a technically optimal solution, and presentation of the proposal to Congress and the public as a finished product.

Regrettably the energy-policy task force failed to meet the political criteria, a failure that was one of the major causes of the political difficulties that the plan's sponsors later encountered. Reflecting a pattern established by other Carter policy-planning initiatives, there was only modest effort to build the political coalition necessary for public approval and for adoption on Capitol Hill. According to the *New York Times*, "The plan was conceived in secrecy by technicians, challenged in haste by economists, and altered belatedly by politicians." This may be somewhat of an exaggeration, but the political ineptitude was evident. There was little direct consultation with Congress, including key committee chairmen, until just before the package was completed. Several of the task-force members were recruited from congressional staffs, and Schlesinger arranged group breakfast meetings at the Capitol and in the White House to explain the plan as it was being drafted. However, the consensus of those involved seems to be that these efforts were insufficient to counteract the earlier lack of consultation.

Congressmen were not the only people whose advice and expertise were not sought by the task force. There was also little involvement by energy company executives, other business people, and environmentalists. Moreover, the draft plan was not circulated widely within the executive branch itself. The energy tax proposals were not checked with the Treasury, the Secretary of Transportation was virtually excluded from the process, and the Office of Management and Budget was consulted only after the basic plan was drafted.[12]

Such an approach may offer important advantages when the goal is no less than a comprehensive package covering all aspects of energy supply and demand: fuels, technology, conservation, taxes, regulation, environmental impact, industrial capacity, imports, and federal-state relations. The final plan, in fact, contained 113 separate and interrelated provisions and when introduced in the House was a 283-page bill. Lack of consultation with other government departments may also have been necessary to save time in meeting the president's deadline. Nevertheless, such a process carries a major political liability: those who are not consulted as a policy package is formulated will have little obligation to support it on the Hill. Given this process of developing the national energy plan, it comes as no surprise that the plan was criticized later for being hastily prepared and poorly conceived. Carter's hope that an aroused public would provide sufficient support for his plan against its congressional and industrial critics proved unrealistic.

Those who were disappointed in not being able to shape policy formulation in the White House turned their attention to Congress. Among those active in opposition on the Hill were energy-company lobbyists, who considered the Carter package deficient in its limited concern for increasing energy production, and consumer and labor representatives, who viewed with disfavor the economic penalities that consumers would have to bear with higher energy prices. In many respects, Congress responded in one of its classic roles.

### *Legitimation? Congressional Action on the Nation Energy Plan*

The legislative history of the national energy plan is both long and complicated, but for present purposes a brief review of the major actions will suffice.[13] The House of Representatives passed the president's bill in substantially the form submitted to Congress, and with uncharacteristic speed. The Senate, on the other hand, rejected most of the key proposals, taking a year and a half to complete its work on the bill.

On May 2, 1977, the majority leader of the House, Jim Wright (D-Texas), introduced H.R. 6831, a comprehensive bill containing President Carter's energy plan. The major provisions of the bill included the following:

A tax credit for home owners installing solar-energy equipment or other devices for conserving energy in their residences.

A gas-guzzler tax designed to penalize owners of inefficient cars and to reward owners of more-efficient cars.

A standby gasoline tax of five cents per gallon to be imposed each year that gasoline consumption did not taper off (with a per-capita rebate provision).

An investment tax credit to provide incentives for businesses to install energy-saving, solar, or cogeneration equipment.

A crude-oil equalization tax designed to raise the price of domestic oil to that of foreign oil and thus discourage consumption (also with a per-capita rebate provision).

A tax on industrial use of oil and natural gas to provide an incentive for conversion to the use of coal (with tax credits for the cost of conversion).

Revision of the system of federal controls on the price of natural gas.

Electric-utility rate reform designed to make rates more accurately reflect the cost of providing consumers with electricity.

Development of plans through which utilities would educate and help consumers make their homes more energy efficient.

Development of energy-efficiency standards for consumer products and procedures for disclosing product compliance with such standards.

A federal grant program designed to encourage hospitals and schools to conserve energy.

A prohibition on the use of oil and natural gas by new fuel-burning plants and requirements for conversion to other fuels for plants now burning oil and gas.

Programs to encourage lending institutions to finance home improvements related to energy savings, weatherization of low-income homes, and development of new building standards to conserve energy.

Establishment of van pooling for federal employees.

Demonstration programs for solar heating and cooling in federal buildings.[14]

The Speaker of the House, Thomas P. O'Neill of Massachusetts, had skillfully arranged for the bill to be referred to five standing committees, with an agreement that the parts would be assembled into a single bill following their action. With a deadline of July 13 imposed, the legislation coming out of the committees was to go to a new House Ad Hoc Select Committee on Energy prior to consideration by the full House. Those parts of the bill dealing with taxation (title II) were referred, as required of all tax legislation, to the Ways and Means Committee. The Interstate and Foreign Commerce Committee received the next largest portion, with smaller segments of the package going to the Banking, Finance and Urban Affairs Committee, the Government Operations Committee, and the Public Works and Transportation Committee. Four of the five committees completed their work on schedule; the Interstate and Foreign Commerce Committee was only one day late.

The chairman of the Ad Hoc Energy Committee, Thomas Ashley (D-Ohio), introduced a new bill (H.R. 8444), which contained the assembled parts handled by the five committees. The Ad Hoc Committee operated as an arm of the Democratic leadership, with the strong support of Speaker O'Neill, and completed its task quickly between July 20 and July 22. This extraordinarily rapid action on so complex a bill was made possible by the composition of the Ad Hoc Committee, as well as by the forceful leadership provided by the Speaker. Most of the members of the committee also sat on the five standing committees that had already reviewed the major sections of the bill. Moreover, O'Neill had personally chosen nearly all of the members of the committee, most of whom were basically sympathetic to Carter's approach. Voting in the committee was along party lines, and since the Democrats outnumbered the Republicans by twenty-seven to thirteen (reflecting party alignment in the full House), Republicans found themselves virtually eliminated from the process. Several amendments were approved by the committee, including a new four cents per gallon federal gasoline tax and an expanded definition of new natural gas, which included more wells than allowed under the Carter plan.[15]

On August 5 the full House passed H.R. 8444 by a vote of 244 to 177. Eighty-two percent of the Democrats voted in favor of the bill, but only 9 percent of the Republicans did so. The new gasoline tax, despite support from both President Carter and the House Democratic leadership, was soundly defeated. Following defeat of a natural-gas deregulation proposal, the House accepted the Select Committee's compromise on the definition of new natural gas.

In contrast to the House, the Senate made no special provisions for handling the energy bill. The Senate majority leader, Robert C. Byrd (D-West Virginia), expressed doubt that the Senate could even complete action in 1977, and he made little effort to persuade it to do so. The energy program was introduced in six bills and referred to the Committee on Finance, chaired by Russell B. Long of Louisiana, and to the new Committee on Energy and Natural Resources, chaired by Henry M. Jackson of Washington. Committee action on the various bills followed no overall strategy or time-table and resulted in legislation containing provisions that departed markedly from the original Carter proposal.[16]

On July 25 the Energy and Natural Resources Committee reported S. 977 on coal conversion by electric utilities and major industrial plants, and the Senate passed the bill on September 8 by a vote of seventy-four to eight. S. 977 was then merged with another bill, which authorized $900 million in federal matching grants for energy conservation in schools and hospitals. On August 18 the Energy and Natural Resources Committee reported S. 2057, which authorized $1.022 billion for a variety of energy conservation programs. The full Senate passed S. 2057 on September 13 by a vote of seventy-eight to four, following three days of debate. These three bills were the least controversial of the six.[17]

On September 15 the Energy and Natural Resources Committee reported S. 2104 on natural-gas deregulation without recommendation because the committee was deadlocked on natural-gas pricing. The Senate debated S. 2104 for fourteen days, including a nine-day filibuster led by Senators James Abourezk (D-South Dakota) and Howard M. Metzenbaum (D-Ohio), who were opposed to natural-gas deregulation, a position favored by a majority of the Senate. The filibuster was finally broken by majority leader Byrd and Vice-President Walter F. Mondale using a controversial procedure to speed action on the energy bill (and in the process irritating senators who objected to such tactics). A modified deregulation substitute offered by James B. Pearson (R-Kansas) and Lloyd Bentsen (D-Texas) was adopted on October 4 by a vote of fifty to forty-six. The bill contained provisions ending federal price regulation for new natural gas found onshore and setting a two-year upper limit on the price of deregulated new natural gas. It also allowed continued regulation of off-shore gas through 1982, defined new gas as that sold or delivered for the first time in interstate commerce after January 1, 1977 (with allowance for specifications of ineligibility by the Federal Energy Regulatory Commission), and established a pricing system for the allocation of old gas (at lower regulated prices) to high-priority users such as schools and hospitals.[18]

On September 19 the Committee on Energy and Natural Resources reported out S. 2114 on utility-rate reform and conservation. However, virtually all of the innovative Carter recommendations were dropped from the bill. The bill's floor manager, Senator J. Bennett Johnston (D-Louisiana), explained that the committee found Carter's initiatives a "radical extension of federal authority into the highly complex matter of the design of retail rates for electricity," a matter he preferred to leave to the states to decide.[19] The weaker bill passed the Senate eight-six to seven.

Carter's energy plan received the least friendly reception of all in the Senate Finance Committee. The committee took a completely different approach to energy tax proposals, favoring tax credits and other incentives to increase production rather than the Carter-proposed tax penalties to reduce consumption, which were to be combined with rebate provisions to return the money to the public. As *Congressional Quarterly* put it, the committee "systematically rejected each of Carter's three key tax proposals—the equalization tax on crude oil, the tax on utility and industrial use of oil and natural gas and the tax on 'gas guzzling' cars; together these taxes accounted for roughly half of the Carter energy program."[20] The committee reported out its bill on October 21, and the Senate approved it, after a six-day debate, on October 31, by a vote of fifty-two to thirty-five. By most accounts, Senator Long dominated the floor debate; the amendments he favored were approved by the full Senate and those he opposed were rejected.[21] Long's success reflected not only his dominance of the committee but the willingness of the Senate to go along with the committee's position on the bill. The president's supporters were unable to persuade a majority of the Senate to support him. These actions in the finance committee and on the floor were significant enough for the press to refer to the net contribution of the Senate as "hacking up," "emasculating," and "devastating" the Carter program.

The House-Senate conference on the energy legislation dragged on until October 12, 1978, nearly a full year after initial Senate approval. The protracted conference negotiations can be attributed to the sharp conflict between House and Senate versions of the bill. Natural-gas pricing and the split between consumer and producer interests lay at the center of this dispute. As expected President Carter and Secretary Schlesinger pressed for resolution of House-Senate differences and for passage of the final legislation, notwithstanding the fact that it fell considerably short of their expectations. The Senate adopted the conference report on October 15, following a fifteen-hour filibuster by the bill's opponents. On the same day, at 7:30 A.M., the House passed the final package, thus ending the lengthy congressional action on the national energy plan. In spite of the obviously limited nature of the legislation, the Carter White House issued a typically upbeat presidential statement upon passage of the bill: "We have declared to ourselves and the world our intent to control our use of energy and thereby to control our own destiny as a nation."[22]

Table 4-1 provides a summary of congressional action on the Carter program. The legislation was not as far-reaching in its scope, not as strong in forcing Americans to curb their prodigious appetite for energy, and not as likely to reduce oil imports as the original proposal. The best that can be said is that it was a modest beginning toward the development of a more expansive and effective national policy in the future.

The failure of Congress to approve the national energy plan or to devise an equally effective national policy of its own is subject to various interpretations, partly reflecting the values of the person evaluating the case. But regardless of personal values, there is some profit in asking why Congress took the action it did.

## Explaining Congressional Policy Making

All case studies are of necessity selective in the determinants of public policy chosen for emphasis. The causes of policy making are too numerous to permit an exhaustive account, and most of us have some preconceived notions about which political, economic, cultural, and social forces are most influential in a given case. I will focus on two broad categories of factors to help explain congressional policy making in 1977-1978: the institutional characteristics of Congress and political variables such as public opinion, interest-group activity, political calculations of members of Congress, and presidential leadership.

### *Institutional Characteristics*

Both the structure of Congress and its style of decision making affected its handling of the Carter plan. Congress is poorly organized for consideration of comprehensive policy initiatives, and its members are increasingly unwilling to follow presidential dictates. However, the combination of congressional organization and political motivation does create numerous opportunities

**Table 4-1**
**Congressional Action on Carter Energy Proposals, 1977**

| *Proposal* | *House Action* | *Senate Action* | *Conference* | *Final Action* |
|---|---|---|---|---|
| Tax credits for home insulation | Approved | Approved | Maximum of $300 credit approved | Approved Oct. 15, 1978 |
| Increase in gasoline tax | Rejected | Rejected by committee | | |
| Tax on gas-guzzler cars | Approved | Rejected; ban on production | Approved | Approved Oct. 15, 1978 |
| Rebate of gas-guzzler tax to buyers of gas-saving cars | Rejected by committee | Not considered | | |
| Mandatory energy-efficiency standards for home appliances | Approved | Approved | Approved | Approved Oct. 15, 1978 |
| Extension of natural-gas price controls, with higher ceilings | Approved | Rejected; ended price controls for new gas | Agreement to end price controls on new gas by 1985 | Approved Oct. 15, 1978 |
| Tax on crude oil | Approved | Rejected by committee | Killed by conference | |
| Tax on utility and industrial use of oil and natural gas | Approved with changes | Approved in weaker version than House | Killed by conference | |
| Coal conversion | Approved | Approved in weaker version than House | Compromise | Approved Oct. 15, 1978 |
| Electric-utility rate reform | Approved | Rejected in committee | Compromise | Approved Oct. 15, 1978 |

Source: Adapted from Congressional Quarterly, *Energy Policy* (Washington, D.C.: Congressional Quarterly, 1979), p. 6.

for dissenting points of view to be heard. Four characteristics significantly affected Congress's legislative actions in this case: (1) the fragmentation of power inherent in its bicameral structure, (2) the dispersal of power within each house, resulting from a system of strong standing committees with overlapping jurisdiction on energy-related issues, (3) weak party leadership in the Senate, and (4) Congress's self-defined legislative role.

The fact that Congress consists of two independent houses, both of which must approve legislation, nearly always complicates the job of policy innovation. Those seeking policy change must win in both houses, but the procedures, internal norms, and political forces within each chamber are quite different. In 1977 Carter's easy victory in the House was largely undone in the Senate, a development apparently unanticipated by the White House. Political naiveté or excessive optimism on the part of White House staff may explain part of the problem, but there were two additional factors: the traditional August congressional recess occurred between House passage of the bill and Senate consideration, slowing the momentum gained from the April presidential address and the activities in the House through late July; and the Bert Lance affair detracted from the attention given to energy issues in late summer.

An even greater problem for the Carter forces was the congressional committee system. The diversity and complexity of modern legislation requires a division of labor and issue specialization represented by the system of standing committees, yet the committee system is very slow to adapt to new issues like energy. By one count in 1976, for example, a total of twenty-three committees and fifty-one subcommittees on the Hill worked directly on energy issues of some sort.[23] In 1979 the staff of the House Select Committee on Committees found that eighty-three subcommittees and committees in the House alone shared in some aspect of energy policy.[24] Strong party leadership can compensate for such overlapping jurisdiction and the tendency to battle over turf, and Speaker O'Neill exercised such leadership in the House. However, majority leader Byrd chose not to do so in the Senate, or he simply recognized the futility of trying had he been so inclined. The Senate generally has not tolerated such leadership since Lyndon Johnson's reign as majority leader in the 1950s. As a consequence of weak party leadership, power over energy legislation in the Senate fell to Henry Jackson and Russell Long. Neither chose to lead the fight for Carter's national energy plan. In fact, Senator Long, a close friend of the oil and natural-gas industry, was instrumental in defeating its key provisions.[25] Carter was left with no strong ally in the Senate to counter these forces. Part of the blame, however, surely lies with the ineffective Carter lobbying effort on the Hill.

The unwillingness of Congress to cooperate with the White House on energy legislation is as much a function of its self-defined legislative role as anything else. Following the experience of Vietnam and Watergate, Congress is more predisposed to assert itself as an independent force in national policy making.[26] The increasing security of incumbents on the Hill probably reduces further their need to cooperate with the president, even when he is of the same party. Moreover Congress now has a greater institutional capability to assert itself with the services provided by the new Congressional Budget Office, the Office of Technology Assessment, the expanded Congressional Research Service, and larger professional staffs attached to its standing committees.

Given these predispositions and capabilities, the drafting of the energy plan in virtual secrecy practically guaranteed congressional opposition. In addition to opposition generated by the president's lack of consultation with Congress, some members were convinced that the plan itself was poorly drafted. In a series of interviews with a cross-section of the Senate in October 1977, *Congressional Quarterly* found that senators generally agreed that the Carter program was "poorly conceived," "was not thought out," and would have "numerous bad effects without saving very much energy."[27] These skeptical views were supported by the findings of four independent analyses by various Capitol Hill research units (the Congressional Budget Office, the Office of Technology Assessment, the General Accounting Office, and the Library of Congress) that Carter's plan would fall short of attaining its energy goals.[28] Studies like these merely emphasized the substantive inadequacies of the president's program, for its political vulnerability was apparent from events on the Hill.

### *Political Conditions*

The political difficulties encountered by the White House may be attributed largely to the general lack of public concern for energy problems and a specific lack of public support for the solutions proposed in the national energy plan. Without both public concern and public support for the plan, the massive lobbying campaign by those opposed was successful. President Carter's weak political leadership could not counter this lobbying effort or build enough public support to save the plan.

It is usually difficult to measure public opinion on policy issues with a high degree of accuracy, and it is even more difficult to show conclusively its impact on policy making. Yet there is little doubt about some basic facts of the public response to the energy crisis. During the period 1977-1978, there was widespread disbelief in the reality of the crisis against which the president was waging his "moral equivalent of war." Moreover the American public was very poorly informed on the facts of the nation's appetite for energy and its dependency on OPEC oil to satisfy that need. Partly as a result of those two conditions, energy policy was not very salient to the average person. The major political consequence was that there was no strong and persistent national constituency for the national energy plan.

A few of the figures are worth reciting. A New York Times/CBS poll conducted in September 1977 showed that only 12 percent of the public thought that a world oil shortage was likely in the next fifteen years. The same poll indicated that in spite of four years of publicity on rising oil imports, one-third of the sample thought that the United States produced all of the oil it requires, and only 48 percent knew that the United States imports oil to meet its energy needs. When asked directly what percentage of the nation's oil was imported, less than one-fourth of the sample came within 10 percent of the correct answer.[29] Gallup polls during 1977 and 1978 indicate that public concern and the saliency of the issue actually declined during the period in which Congress was considering Carter's plan. In July 1977 15 percent of a Gallup sample cited energy as the "most important problem" facing the nation. By October 1978 only 3 percent rated the energy problem as most important. The percentage citing energy as a very serious problem did not change at all in 1977-1978, despite all of the attention in the media and the emphasis given the issue by the president.[30]

In April 1979 a New York Times/CBS survey found that only 33 percent of the public believed that energy problems "were as bad as the President said," representing no change at all from July 1977.[31] Two years of presidential effort to build public concern for energy scarcity seemingly had failed. Extensive media speculation on the artificiality of energy shortages and criticism of windfall oil-company profits certainly did not make the president's job easier, nor did the generalized public cynicism and distrust of government that characterized the late 1970s. One is tempted to say that President Carter might have countered public skepticism to some extent by addressing more effectively the issue of energy-company monopoly. Widespread public suspicion of the power of oil companies to control energy supplies and prices seems to have been a major contributing cause of the rather weak public support for the energy plan itself. Public apathy and distrust continued through 1979, and by May of that year, a frustrated president had taken to scolding the American people for their unconcern and accusing the House of Representatives of "a remarkable demonstration of political timidity" for rejecting his proposed plan for emergency gasoline rationing.[32]

Survey data of this type are hardly conclusive. The public's willingness to support energy-policy initiatives depends upon its knowledge of the facts and its belief in the need to act, both of which can be affected by political leadership, coverage of the issues in the mass media, and events such as the Three Mile Island nuclear accident in 1979. But during the 1977-1978 period in which the president's energy plan was under consideration on the Hill, public consensus on the program was lacking, and the public was highly distrustful of the type of governmental proposals being made. By rejecting the plan in its initial form, Congress was accurately reflecting the public's distrust and its reservations on national energy policy.

The lack of public consensus on energy policy and the absence of a large constituency for the Carter program left it vulnerable to organized pressure. The national energy plan was the object of intense, persistent, and largely successful lobbying by a wide range of interest groups opposed to its provisions. A labyrinthine legislative process on Capitol Hill in combination with such group activity normally favors outcomes preserving the status quo.[33] This is especially likely when business or industry groups, generally better organized and financed and having greater access, predominate or when there is no effective representation of the larger public interest.[34]

The major pressures in 1977-1978 came from the oil- and natural-gas producing industries seeking additional financial incentives for energy production, utilities opposed to Carter's utility rate reform and coal-conversion programs, automobile companies unhappy with the tax on gas-guzzling cars, and labor, consumer, and environmental groups with various objections to the impact of rising energy costs on consumers and the environmental consequences of particular forms of energy use. *Congressional Quarterly* aptly described their activities: "The Capitol was crawling with lobbyists of every shape, stripe, and persuasion from the day Carter sent his energy package to Congress."[35] The number and diversity of groups does not imply, of course, that each was equally influential or that the public interest was necessarily well served by the competitive group process, especially in Senate action on the bill. Energy is such a multifaceted policy area that even so-called public-interest lobbies have staked out quite disparate positions.[36] But although no particular policy position can be objectively identified with the public interest, one must wonder whether the political process as a whole failed to protect the nation's collective and long-term interest. Not surprisingly, President Carter has frequently expressed this view, notably in his July 15, 1979 speech on the crisis of confidence and governance in the United States.

The difficulty of cooperative effort on behalf of national energy policy can be understood by appreciating the political constraints within the legislative process. Congress is composed of 535 politicians whose electoral security is always in some doubt and who therefore believe they must be responsive to public opinion and its expression through organized groups. Elected from local and state, not national constituencies, they are particularly responsive to local and regional interests and are concerned with local consequences of national policy actions.[37] Moreover they are likely to identify policy problems in a short-term rather than long-term context, are not likely to be well informed on the technical aspects of the issues, and are not ordinarily predisposed toward energy conservation and ecological concerns when they are thought to affect economic growth and public convenience adversely.[38] These political incentives and attitudes add up to considerable trouble for a president who is proposing national energy policies that emphasize short-term, highly visible, and personal sacrifices for long-term and diffuse collective benefits.

Presidential leadership sometimes makes the task of consensus building and policy making possible in spite of these constraints. However, President Carter was not successful in that regard in 1977-1978, and he continued to demonstrate weak leadership on energy issues in 1979. Carter's problems began with the process of policy formulation; the energy-policy task force seriously misjudged the plan's acceptability. But the major difficulties flowed from his inability throughout the two-year period to use the resources of the presidency fully to sell the plan to the American public and to Congress. To be sure, there was an attempt to do so. Administration officials, supplied with pamphlets and question-and-answer sheets, appeared on television talk shows, gave speeches, and granted interviews; hundreds of thousands of booklets explaining the energy plan were distributed to opinion leaders around the country; the Democratic National Committee was enlisted to sell the program to state and local party leaders; and Secretary of Energy Schlesinger and his staff conducted extensive briefings for the press, members of Congress, and business executives, both in Washington and around the country.[39] But the impact of this activity was less than overwhelming.

The poor results can be traced to Carter's inexperience in policy formulation and public relations, his unfamiliarity with Washington ways, a variety of political misjudgments in dealing with Congress, and the normal constraints on presidential decision making due to limited time, attention, and resources. A few examples illustrate these problems. To begin with, there was a lack of follow-through between the April speech and late September, when the president attacked the deregulation forces in the Senate for having devastated his plan. For an issue said to be one of his top priorities, the inactivity is puzzling. At least part of the explanation, however, is the inability of any president to devote as much time and attention to some issues as he would like. In President Carter's case, his limited time and staff resources were drained away by major foreign-policy issues (especially SALT II and Middle East controversies), a deteriorating economy at home, and other domestic-policy problems.

The president also failed to generate enough public concern to bring political pressure to bear on Congress. Within a few days of his first energy speech on April 18, 1977, Carter deemphasized the sacrifices he initially called for, saying that his program would have "no significant effect" on the nation's economic growth or standard of living.[40] In what may have been an attempt to present his program in a more positive light, Carter created the appearance of inconsistency, which further weakened his already limited credibility on energy issues. Perhaps a more forceful and consistent explanation of long-term energy goals and necessary changes in personal life-styles, in the economy, and in transportation, housing, and other policy areas would have been more effective in building the constituency for his energy program that was so obviously lacking.[41]

The administration's lobbying effort on the Hill also was notably unsuccessful. The Carter White House was inconsistent and ambivalent on some of the key issues; much of the lobbying was carried out by inexperienced and poorly informed aides; there was a severe underestimation of the opposition forces, especially in the Senate; and the president himself was less personally involved than one might have expected from his public statements on energy. Carter's critical attitude toward Congress won him few friends on the Hill, and his predominant style of leadership made a cooperative and productive relationship with Congress unlikely.

**Conclusions**

In many respects the refusal of the Ninety-fifth Congress to approve President Carter's national energy plan demonstrates the normal workings of the American political system. The president formulated and sent to Congress a reasonably comprehensive and rational package of energy-policy proposals designed to reduce the nation's consumption of oil and natural gas and to limit oil imports. Acting in its legitmate policy-making capacity, Congress defined energy problems differently and substituted its own judgment for the president's. In doing so it made a series of incremental adjustments to existing policies that reflected public uncertainty over the seriousness of the energy crisis and a lack of consensus on the best solution. Thus no national policy emerged, and some of the most complex and difficult issues were left for future action. That result led some to question the capacity of the federal government (in particular the Congress) to make and implement comprehensive energy policy, caused others to wonder whether the public interest is necessarily well served by a comprehensive and coherent policy at this time or by executive dominance of policy making on these issues, and prompted still others to propose a range of research and political strategies that might improve energy policy making in the future.

Those who favor comprehensive policy planning are generally disappointed with the failure of Congress to approve the Carter plan. Their reviews of Congress's action tend to emphasize that institution's parochialism in favoring state and local interests over the national interest, the political timidity of its members in the face of public reluctance to pay higher energy prices and to conserve energy resources, and its determinedly short-term outlook on a long-term problem. Congressional incapacity to make policy in a coordinated and comprehensive fashion is especially singled out for criticism, as is its vulnerability to the pressures of organized interests.

Some environmentalists carry the argument further. Many consider the national energy plan to have been something less than comprehensive and less than rational from the perspective of environmental costs of energy production and use and the weak support given to the use of renewable energy resources. Some, most notably Barry Commoner, thought that the president failed to challenge the concentration of economic power represented by the major energy companies. And some of the more theoretically minded environmentalists express little confidence in the capacity of the American political system to resolve energy issues. They point to such obstacles to energy policy making as the pluralistic nature of American politics, the fragmentation of power in Congress, the power of corporate interests, a capitalist economic system, and a political culture promoting individual self-interest over the national interest. Solutions offered tend to go beyond particular policy proposals to moderate and more-radical suggestions for political change: extensive reforms designed to strengthen public-interest politics and the role of policy planning and radical changes in government, politics, the economy, and life-styles, with some curtailing of democratic processes and individual rights and greater reliance on technical experts and planners in decision making.[42]

A quite different assessment of the energy-policy process is offered by pluralists and others partial to an active and independent Congress. They argue that energy issues are heavily political in nature and therefore should be acted on by democratically accountable legislators, not solely by policy analysts, economists, and planners. In their view, Congress is likely both to consider questions neglected by the executive branch and to take a distinctly different position on the issues.[43] The social and economic costs of policy proposals are examined more thoroughly, and numerous opportunities are created for various organized groups to present their views. Although the large number of individuals involved in this deliberative process hardly guarantees speedy or coherent action, it does promote resolution of the issues through bargaining and compromise and therefore tends to build public consensus. The result is often greater public support for policy decisions and more effective policy implementation. Pluralists also argue that the lengthy process of developing national policy allows considerable innovation and experimentation at state and local levels, which may significantly influence national policy decisions.[44]

My own view is that one need not claim special powers for definining the public interest in these matters to suggest that the nation must at least choose the energy future it wants by adopting a national policy that sets forth both short-term and long-term goals and priorities and establishes programs likely to achieve those goals with minimal sacrifice of other important values. The alternative of muddling through without such a policy involves unacceptably high costs, whether viewed in terms of national security, economics, environmental impact, or public health and safety.

One way to appreciate those costs is to compare current issues in energy policy (which are concerned chiefly with energy availability, economic costs, and national security implications) with the broader, long-range social, economic, and political issues related to a transition from resource abundance to the scarcity associated with ecological limits to growth. The United States (and all other nations) must eventually adjust to resource scarcity and turn to sustainable sources of energy like solar power, and the costs of such a transition will be substantially greater without the foresight and planning that would accompany the establishment of national policy goals and programs.[45] The adoption of such a national policy presumably would follow careful and thorough analysis and widespread debate on the alternatives and their consequences. And a national policy need not, of course, unduly constrain state and local policy innovation or personal and private-sector initiatives.

Whatever one's assessment of the events surrounding the national energy plan, there is probably agreement that the future of energy policy depends upon three conditions: (1) clarifying the type of energy future and paths of development that may be said to represent the public interest; (2) building an understanding of the technical and economic problems of energy development and its consequences (especially national security, environmental, economic, and social impacts of energy-use patterns), and (3) building public support sufficient to enact and implement effectively public policies that will guide energy development toward the type of energy future that the nation determines is in its long-term interest. None of the three conditions will be achieved easily, but at least the outlook seems to be improving.

The late 1970s saw a proliferation of energy-policy analyses that addressed the questions of energy development and its impact. In contrast to studies of several years ago, the technical quality of the major works of the past year or so, as well as their objectivity in evaluating alternative energy policies, is quite high. Partly for those reasons, the current studies seem to be well received by policy makers of diverse political philosophies and interests; a consensus on technically and economically feasible energy policies may be emerging slowly. For example, a number of new studies appearing in 1979 agreed that the conservation of energy (especially "wasted" energy) can make a major contribution to the nation's short-term energy needs, that economic growth need not suffer as a result of conservation policies, that some short-term environmental tradeoffs probably will be necessary, and that long-range needs should be met with renewable-energy sources. At the same time, nuclear power is viewed as unlikely to make a major contribution to U.S. energy sources in this century, primarily because of technical and political problems. Most of these studies agree that decontrol of domestic oil and natural-gas prices is essential in the near future and that with "proper policy and planning and a willingness to pay the costs," the United States will not run short of energy.[46]

Development of a national consensus on what those policies should be and what costs will be borne by whom necessarily depends upon leadership and public education on the issues. Such a development will come only slowly and will depend upon more than the production of good energy-policy analyses. The analyses must be persuasive to the public and its political representatives. Issues such as energy costs and social equity and environmental-energy trade-offs will not be resolved without considerable controversy, and presidents and other policy makers attempting to devise solutions will find the building of public consensus difficult. Moreover the 1980s and 1990s may well see the continuation of generalized public cynicism toward government and politics and skepticism toward various claims for the beneficent effects of new programs said to advance the national interest. Yet some modest signs show that the nation slowly is becoming more informed about energy production and use. Reaction to the Three Mile Island accident in early 1979 and concern over the economic impact of our continuing reliance on OPEC oil are indicative of changing public attitudes. Innovative policy developments in selected states and localities and private-sector and personal decision making on energy conservation are further positive indicators.

There is, in short, no fully convincing set of reasons to suggest an optimistic attitude toward the development of public support for a rational energy future for the United States. Yet one need not adopt an entirely pessimistic stance either. Perhaps a position of hopeful pessimism is in order. Energy problems need to be appraised with the seriousness they merit, and the kind of political and institutional constraints I have identified need to be borne in mind as public policies are formulated and political strategies are devised. Those tasks may suggest an attitude closer to pessimism than optimism. However, one needs to be hopeful that solutions can be found and that active political concern and participation will indeed make a difference in determining the nation's energy future. The way in which the nation copes with its energy problems will be shaped by those people who are sufficiently concerned about the issues to play active roles in research, analysis, public debate, and decision making.

## Notes

1. *New York Times,* July 17, 1979.

2. *Newsweek*, July 30, 1979. The Kennedy-Durkin proposal was based in part on a well-received study by the Harvard Business School: Robert Stobaugh and Daniel Yergin, eds., *Energy Future: Managing and Mismanaging the Transition* (New York: Random House, 1979).

3. Charles O. Jones, *An Introduction to the Study of Public Policy,* 2d ed. (North Scituate, Mass.: Duxbury, 1977).

4. David Howard Davis, *Energy Politics,* 2d ed. (New York: St. Martin's, 1978), and "Energy Policy and the Ford Administration: The First Year," in David A. Caputo, ed., *The Politics of Policy Making in America* (San Francisco: W.H. Freeman, 1977), pp. 39-70.

5. Congressional Quarterly, *Energy Policy* (Washington, D.C.: Congressional Quarterly, 1979), p. 9-A.

6. Ibid., p. 36-A.

7. Ford Foundation Energy Project, *A Time to Choose: America's Energy Future* (Cambridge, Mass.: Ballinger, 1974). The criticisms can be found in Institute for Contemporary Studies, ed., *No Time to Confuse* (San Francisco: Institute for Contemporary Studies, 1975).

8. David Howard Davis, "America's Non-Policy for Energy," in Theodore J. Lowi and Alan Stone, eds., *Nationalizing Government: Public Policies in America* (Beverly Hills, Calif.: Sage, 1979), pp. 101-128.

9. Congressional Quarterly, *Energy Policy,* p. 5-A.

10. *New York Times,* April 19, 1977.

11. Executive Office of the President, Energy Policy and Planning, *The National Energy Plan* (Washington, D.C.: Government Printing Office, 1977).

12. Descriptions of the energy task force and its work are taken from *New York Times,* April 10, 24, 1977, and from Congressional Quarterly, *Energy Policy,* p. 34.

13. A full review of the legislative actions in the 1977-1978 case can be found in the excellent compilation prepared by Congressional Quarterly, *Energy Policy.* The legislative history presented here relies heavily on that thorough report. Those wishing even more detailed information on congressional policy making should consult committee hearings, committee reports, and the *Congressional Record.*

14. Congressional Quarterly, *Energy Policy,* pp. 39-40.

15. Ibid., pp. 40-41.

16. Ibid., pp. 41-44.

17. Ibid., pp. 41-42.

18. Ibid., pp. 42-44.

19. Ibid., p. 44.

20. Ibid.

21. Ibid.

22. Ibid., p. 5.

23. Jones, *Introduction to the Study of Public Policy*, p. 75.

24. *Congressional Quarterly Weekly Report*, November 3, 1979, p. 2486.

25. *New York Times*, April 15, 1979.

26. With respect to legislative roles, Charles O. Jones argues that there are three views of the relationship of Congress and the president in energy policy-making responsibilities: the president can dominate in policy formulation and expect Congress to be largely a silent partner; the Congress itself can be the initiator of comprehensive energy policies, challenging the president or acting in the absence of presidential initiatives; or Congress can be a facilitator, representing pluralistic interests and modifying presidential proposals to find a legitimate or acceptable compromise. See Jones, "Congress and the Making of Energy Policy," in Robert Lawrence, ed., *New Dimensions to Energy Policy* (Lexington, Mass.: Lexington Books, D.C. Heath, 1979), pp. 161-178.

27. *Congressional Quarterly Weekly Report,* October 22, 1977, p. 2233.

28. Congressional Quarterly, *Energy Policy,* p. 34.

29. *New York Times,* September 1, 1977.

30. *Gallup Opinion Index* (May 1977), (October 1978).

31. *New York Times,* April 15, 1979.

32. Ibid., May 15, 1979.

33. David B. Truman, *The Governmental Process* (New York: Alfred A. Knopf, 1951).

34. E.E. Schattschneider, *The Semisovereign People* (New York: Holt, Rinehart, and Winston, 1960); Theodore J. Lowi, *The End of Liberalism*, 2d ed. (New York: Norton, 1979); Mancur Olson, *The Logic of Collective Action* (New York: Schocken, 1965).

35. Congressional Quarterly, *Energy Policy,* p. 35.

36. Andrew S. McFarland, *Public Interest Lobbies: Decision Making on Energy* (Washington, D.C.: American Enterprise Institute, 1976).

37. David R. Mayhew, *Congress: The Electoral Connection* (New Haven: Yale University Press, 1974); John W. Kingdon, *Congressmen's Voting Decisions* (New York: Harper and Row, 1973).

38. Michael E. Kraft, "Congressional Attitudes toward the Environment: Attention and Issue-Orientation in Ecological Politics" (Ph.D. diss., Yale University, 1973).

39. *Congressional Quarterly Weekly Report*, May 7, 1977, pp. 839-841.

40. *New York Times*, May 19, 1979.

41. The critique presented here has been extended by some journalists to the entire Carter presidency. See, for example, James Fallows, "The Passionless Presidency," *Atlantic Monthly* (May 1979): 33-46, and Tad Szulc, "Our Most Ineffectual President," *Saturday Review*, April 29, 1978, pp. 10-15. The classic account of the effective exercise of presidential power offers a relevant contrast, see Richard Neustadt, *Presidential Power* (New York: Wiley, 1960).

42. For an extended critique of the limitations of the American political process in dealing with energy and environmental problems, see William Ophuls, *Ecology and the Politics of Scarcity* (San Francisco: W.H. Freeman, 1977). A specific attack on both the Carter presidency and Congress for its handling of the energy issue can be found in Barry Commoner's *The Politics of Energy* (New York: Alfred A. Knopf, 1979).

43. Jones, "Congress and the Making of Energy Policy"; Alfred R. Light, "The National Energy Plan and Congress," in Lawrence, *New Dimensions,* pp. 179-190. For a review of the problems of comprehensive policy making and implementation under present political conditions and with the present weak administrative capacities of the executive branch, see Richard J. Tobin and Steven A. Cohen, "The Formulation and Implementation of Energy Policies," in Michael Steinman, ed., *Energy and Environmental Issues: The Making and Implementation of Public Policies* (Lexington, Mass.: Lexington

Books, D.C. Heath, 1979), pp. 173-189, and Terry D. Edgmon, "Organizing for Energy Policy and Administration," in Lawrence, *New Dimensions,* pp. 81-92.

44. The absence of a national energy policy seems to have spurred some local governments (for example, Portland, Oregon, and Davis, California) to develop imaginative and effective conservation and mass-transportation policies. For a review of state energy policy initiatives, see Patricia K. Freeman, "The States' Response to the Energy Crisis: An Analysis of Innovation," in Lawrence, ed., *New Dimensions,* pp. 201-207.

45. See Ophuls, *Ecology;* Dennis C. Pirages, ed., *The Sustainable Society: Implications for Limited Growth* (New York: Praeger, 1977); Commoner, *Politics of Energy.*

46. Robert Stobaugh and Daniel Yergin, eds., *Energy Future: Managing and Mismanaging the Transition* (New York: Random House, 1979); Sam Schurr, Joel Darmstedter, Harry Perry, William Ramsay, and Milton Russell, *Energy in America's Future: The Choices Before Us* (Baltimore: Johns Hopkins University Press, 1979); Hans H. Landsberg, ed., *Energy: The Next Twenty Years* (Cambridge, Mass.: Ballinger, 1979); and Amory B. Lovins, *Soft Energy Paths: Toward a Durable Peace* (Cambridge, Mass.: Ballinger, 1977). See also the report by the Council on Environmental Quality, *The Good News about Energy* (Washington, D.C.: Government Printing Office, 1979).

# 5 Notes from No Man's Land: The Politics and Ecology of Energy Research and Development

*Walter A. Rosenbaum*

In mid-1977, the Reverend Scott Rawlings lifted his voice out of no man's land and prayed. While the governor and five hundred citizens of Portsmouth, Ohio, silently concurred, Reverend Rawlings implored, "We have to pray for our President . . . that he will listen to the voice of God and do what is right." What the supplicants dearly wished was that the Almighty would persuade Jimmy Carter that it was morally imperative to construct a $4.4 billion nuclear fuel enrichment facility in economically depressed Portsmouth.

But if the Almighty turned his ear unto Portsmouth, he would also be turning his face against Oak Ridge, Tennessee, where the facility and its twenty-five hundred permanent jobs were also expected. Between Oak Ridge, Portsmouth, and the Almighty stood Jimmy Carter, wrestling with second thoughts about his presidential campaign promise to move the facility from Tennessee to Ohio. The issue had escalated into a geographical power struggle; into the fray came both state congressional delegations, both community mayors, lobbyists for contractors, energy officials—partisans all—determined to move the facility north or south. In the end Portsmouth prevailed, but the significance of the struggle far exceeds the economic repercussions to either community.[1] The conflict announced that federal spending on energy research and development, a relatively minor budget item a decade earlier, had rapidly become a major economic program and a highly politicized affair. Both events have important implications to environmental management.

During 1980 the federal government will spend an estimated $5.4 billion on energy research and development. Within this decade federal energy R&D spending has sharply risen until it is second only to defense expenditures among all categories of federal R&D obligations.[2] Energy R&D today is roughly equivalent in magnitude to the nation's agriculture program and exceeds any category of environmental expenditure. During the next decade, federal R&D expenditures are estimated to range between $44 billion and $66 billion.[3] Willis Shapley, writing for the American Association for the Advancement of Science,

---

I wish to express my appreciation to the Woodrow Wilson International Center for Scholars, the Smithsonian Institution, for its support during the research and writing of this chapter.

calls major aspects of this R&D program a "no man's land" where a "policy vacuum" invites political resolution, a situation conducive to political bargaining and conflict akin to the Oak Ridge-Portsmouth affair.[4] In a broader sense, energy R&D has been a no man's land because the explosive growth of federal energy appropriations has outdistanced our understanding of the institutions and processes that are shaping this policy. A careful look at the policy structures involved in this R&D is particularly important for anyone concerned with the nation's environmental planning. There are direct ecological impacts in the activities underwritten by this spending and pervasive ecological biases in the policy structures that define substantive goals and program priorities within the energy R&D budget.

Energy R&D is environmental policy by another name. The nation's environmental quality will be affected directly by the mix of technologies promoted through energy R&D. It will powerfully affect the future balance between centralized and decentralized power-generating systems, between resource-conserving and resource-depleting technologies (for example, solar-energy systems versus synthetic-fuel facilities), between environmentally benign technologies (solar and geothermal, for instance) and ecologically hazardous technologies (nuclear versus synthetic fuel). Further, federal energy R&D is the financial life-support system for emerging, environmentally risky technologies whose development, if only to the demonstration phase, would be unlikely without federal participation. The Congressional Research Service (CRS) has noted that the absence of federal R&D support for synthetic-fuel research would mean that "commercial production of synthetic liquids from coal is projected to be zero through 1985 and possibly until 1990 or beyond."[5] The development of energy technologies is emerging as a new, and perhaps unrecognized, institutional procedure for determining the national trade-offs between future energy production and environmental protection.

New policies, as Hugh Heclo remarks, create new politics. Federal spending for energy R&D has created not only new commitments to technology development but a new political structure organized about the management of this spending. This is not itself remarkable. "In general," writes Edward Tufte about governmental management of the economy, "the more important the agency or issue, the more likely and intense the political control."[6] We can expect this political management to become more structured and salient within government as the magnitude of federal R&D spending grows. More significantly, political control implies a particular logic. Like other forms of federal spending, R&D will be treated by the White House and Congress as a political resource—political capital in the business of electoral, partisan, or legislative politics. Thus spending items become a coin in political exchange between actors in the governmental process. Additionally Congress and the White House are likely to compete for control of energy R&D. Congress, in the past, has shown an inclination to insulate such expenditures from direct White House manipulation.

(Significantly, more than one-quarter of all uncontrollable spending in recent federal budgets—spending not alterable by a president within a specific fiscal year—has been spent on federal R&D.)[7] Congress, in short, is likely to exert especially strong influence on this budget, and, consequently, the logic of legislative politics will heavily color spending decisions. Finally whenever Washington sows billions in new expenditures, it reaps new economic dependencies: interests nourished by infusions of federal dollars and mobilized to keep the money flowing. These groups form a cluster of political infrastructures about spending programs; they promote program continuity and content. The effect in all of this is a congeniality to the environmentally hazardous technologies.

### Where the Money Goes

During fiscal year 1980, the federal government will spend an estimated 17 percent of all R&D expenditures on energy, the largest category of civilian R&D spending. The most important environmental aspects of this program are the nature and priority among the supported technologies, the magnitude of funding, and the extent of federal involvement compared with the private sector.

### Supply before Conservation

Federal spending for energy technology development is customarily aggregated under five broad categories. In table 5-1 these categories are decomposed into their major technology components. Generally these categories can also be

**Table 5-1**
**Major Categories and Components of Federal Research and Development Expenditures**

| | |
|---|---|
| Nuclear | Liquid metal fast-breeder reactor, uranium-enrichment process, commercial waste management, spent-fuel disposition, light-water reactor facilities, magnetic fusion, advanced nuclear systems |
| Fossil | Coal liquefaction, coal gasification, oil-shale conversion, fluidized-bed combustion, magnetohydrodynamics, demonstration plants |
| Solar | Biomass conversion, photovoltaic conversion, solar thermal power systems, wind energy conversion, ocean thermal energy conversion, solar heating and cooling of buildings |
| Geothermal | Conventional mining (deep and surface), in situ mining |
| Conservation | Appropriate technology, energy-impact assistance, energy extension service, inventors' program, buildings and community systems, state energy management and planning |

arranged into those programs oriented toward high-volume production with significant ecological hazards during development or operation (fossil and nuclear) and those involving moderate energy production with relatively few adverse environmental impacts (solar, geothermal, conservation). In table 5-2 federal spending on these five categories between FY 1974 and FY 1980 has been summarized by dollar amounts and relative shares of the total energy R&D budget.

Several significant patterns are evident in table 5-2. The most consistent feature is the massive dollar commitment to promoting energy supply. In FY 1974, roughly nine of every ten federal dollars were invested in supply R&D; in FY 1980 approximately two of every three dollars will be spent on nuclear and fossil-fuel systems, both intended to produce large, centralized energy-generating facilities. Additionally the federal government continues to commit by far the largest portion of its energy research resources to nuclear energy. Although nuclear technologies no longer command the almost exclusive share of the R&D budget they did in FY 1974, they will still claim 41 percent of all energy R&D spending in FY 1980. A major change in funding patterns is the rapidly growing importance of fossil-fuel technologies. These technologies —primarily coal liquefaction, coal gasification, fluidized-bed combustion, and oil-shale conversion—are ultimately expected to produce large volumes of synthetic oil and gas. A second change is the progressively larger, though still moderate, share of the R&D budget allocated recently for solar-energy research and energy conservation. During FY 1979 and 1980, these categories will account for about a third of the total energy R&D budget.

From an environmental viewpoint, this rising investment in solar energy and conservation is welcome; however, the balance is still tipped heavily toward the more environmentally ominous technologies. The ecological problems with nuclear-energy technology are well recognized. Synthetic fuels are another matter, for the environmental implications involved in synthetic-fuel production—the fossil category in table 5-2 are less widely understood.

## Synthetic Fuels and Environmental Risks

Enormous uncertainty clouds any discussion of environmental risks in synthetic-fuel technology; U.S. experience with synthetic-fuel production is quite limited, and consequently data are scarce. Existing data, some of it theoretical, strongly warn of potentially grave environmental liabilities. Yet control technologies or newer processes may emerge to reduce the risks to acceptability. Sufficient environmental hazards prevail, in any case, to justify a prudent, searching, and open public assessment of this risk before the nation plunges into an ambitious synthetic-fuel production program.

Coal liquefaction and gasification account for more than half of the fossil-

**Table 5-2**
**Federal Budget Obligations for Energy Research and Development, by Major Categories**
(*$ million*)

| | Fiscal Year | | | | | | | | | | | | | |
|---|---|---|---|---|---|---|---|---|---|---|---|---|---|---|
| *Category* | *1974* | | *1975* | | *1976* | | *1977* | | *1978* | | *1979* | | *1980*[a] | |
| Nuclear | 1,034 | (80) | 1,193 | (65) | 753 | (59) | 2,095 | (68) | 1,378 | (54) | 1,452 | (43) | 1,401 | (41) |
| Fossil | 110 | (9) | 435 | (24) | 323 | (25) | 503 | (16) | 684 | (27) | 826 | (25) | 796 | (23) |
| Solar | 45 | (3) | 102 | (6) | 93 | (7) | 275 | (9) | 368 | (14) | 456 | (14) | 597 | (17) |
| Geothermal | | | 20 | (1) | 31 | (2) | 52 | (2) | 106 | (4) | 136 | (4) | 111 | (3) |
| Conservation | 105 | (8) | 86 | (5) | 66 | (5) | 139 | (5) | 31 | (1) | 504 | (15) | 555 | (16) |
| Total | 1,294 | (100) | 1,836 | (100) | 1,266 | (100) | 3,064 | (100) | 2,567 | (100) | 3,374 | (100) | 3,460 | (100) |

Source: U.S. Office of Management and Budget, *The Budget of the United States*, various years, and *Special Analysis of the Budget*, various years.

Note: Numbers in parentheses are percentages of total.

[a]Estimated in American Association for the Advancement of Science, *Intersociety Preliminary Analysis of R&D in the FY 1980 Budget* (Washington, D.C.: AAAS, 1979)

fuel research budget. Both processes, if commercially viable, will vastly increase the national coal consumption and thus will intensify strip-mine coal demand. A single high-Btu coal-gasification facility, for instance, will require an estimated 6.6 million tons of coal annually. An H-coal liquefaction plant would need about 12.5 million tons of coal yearly.[8] Synthetic-fuel facilities are also likely to create high resource demands at their sites and to generate a variety of hazardous effluents. A commercial HYGAS plant, producing coal gas, may require as much as 1.2 billion gallons of water per hour.[9] Considerable environmental contamination may also be involved in the solid, liquid, and gaseous wastes produced by synthetic production processes. Apprehension is compounded by ignorance because the nature of potential toxics that will be present in plant wastes is often unknown. These themes suffuse a recent Department of Energy (DOE) assessment of environmental impacts from commercial coal gasification:

> An important concern raised by coal gasification processes is the large quantity of solid wastes which will be generated. Currently, little is known about the chemistry or biology of these materials; the toxicity and fate of these solid wastes must be determined.
>
> A major concern relating to commercial coal gasification is exposure of workers and nearby residents to effluents which may contain carcinogenic polycyclic aromatic hydrocarbons (PAH). The potential for acute exposures to carbon monoxide, sulfur oxides and nitrogen oxides is present also.
>
> Quench and clean up waters contain condensed volatile organic and inorganic materials, include regulated and unregulated hazardous agents. . . . Little is known about control of carcinogens that could be regulated . . . or how industry can cope with handling the volumes of liquid waste which might require special handling and disposal.[10]

The department's coal-liquefaction assessment, reciting a similar litany of ecological liabilities, concludes by describing a malevolent chemical cocktail likely to appear as "leachate" from solid wastes:

> It is highly likely that a number of process residues and waste materials will be found to contain toxic, mutogenic, and carcinogenic materials and [to] pose work force, occupational and ecological risk. . . . Process solid wastes from all sources may contain materials which can be leached from the residue, modified by factors in soil and water in ways which change their bioavailability and toxicity, and which will then pose unexplored [hazards to] environmental health.[11]

Among the remaining synthetic-fuel technologies, oil-shale conversion (currently accounting for 6 percent of expenditures) also poses environmental problems. Indeed the oil-shale program's future is precarious because, as the CRS notes, such conversion creates huge amounts of spent shale, mining waste products, water consumption, water and air effluents, and other land impacts.[12] (Consequently it is fiercely regulated; more than one hundred federal state and local permits are required of process developers.)

The full environmental impact of federal energy R&D priorities does not arise from funding levels alone. The extent to which federal participation underwrites the technology development—that is, the relative public and private shares of R&D—bears scrutiny also.

## What Uncle Sam Wants, Uncle Sam Buys

The federal share of energy R&D affects the environment in several ways. The extent of federal funding suggests the extent to which federal decisions—and political structures affecting those decisions—can significantly modify, and perhaps drastically alter, existing research patterns with their differing environmental impacts. Moreover, the magnitude of private-sector dependence on federal funding for R&D activities suggests the intensity and location of pressure for continued program enhancement; in effect, it identifies the stakeholders in present development scenarios.

The federal government heavily underwrites all forms of national R&D; energy is no exception. About half of the nation's total R&D expenditures originate in Washington. Colleges and universities, for instance, receive two-thirds of all funded R&D dollars from federal sources.[13] Energy is among the most heavily subsidized sectors; in 1978 about 35 percent of all energy R&D was funded by industry and the remainder by the federal government.

The federal government's massive involvement in nuclear technology development, the first and only extensive experience with federal investment in energy R&D, created a major energy-policy subsystem linking public and private institutions in the aggressive promotion of expansive R&D funding for the industry. The story is oft told. Suffice it to note that the government's investment of more than $13 billion in civilian nuclear power since 1945 concurrently produced (1) heavy federal subsidization of the industry's development through huge technical and capital resource investments, in keeping with the Atomic Energy Commission's (AEC) commitment to carry technical developments "to the point . . . where the industry can provide nuclear power installations with overall attractiveness sufficient to induce public and private utilities to install them at their own expense";[14] (2) a new political infrastructure—the Joint Congressional Committee on Atomic Energy, the AEC, the nuclear-power lobby, public officials from constituencies affected by federal spending, and academic and professional groups involved with nuclear science—all committed to keeping the program afloat; and (3) many congressional and White House actions creating legal relief from developmental risks. The Price-Anderson Act (1957), for instance, limited nuclear plant operators to a maximum $560 million insurance liability for a nuclear accident and thereby encouraged capital flows into the industry. In these respects, the nuclear program resembled any other Washington program in its tendency to assume an institutionalized form, to create its own political support structure, and to strive for immortality.

The federal fossil-fuel program is already revealing a pattern of heavy subsidization and large dependencies reminiscent of the early nuclear program. Growing by large increments in recent years, the program in FY 1980 will probably exceed $750 million and involve about 25 percent of the energy R&D budget. Washington is heavily committed to subsidizing the development of this technology at every stage short of commercialization. The DOE is committed to funding a third of pilot plant construction and operation costs, half of demonstration facility costs, and the full expense of initial research stages, as well as all pilot and demonstration plant design. The department also reserves to itself generous discretion to exceed the apparent limit of 50 percent funding for demonstration facilities. The General Accounting Office (GAO) explains:

> . . . in the proposal evaluation, a cost/price analysis is included to compute the cost of the proposed contract to the Government. This cost is compared to the benefits before awarding the contract. . . . If a proposal is clearly superior on its technical and economic merit, the cost to the government becomes a secondary consideration, and the agency could accept a proposed project with more or less than 50 percent cost-sharing amounts.[15]

The department will certainly obligate itself to the limits of its authority in synthetic-fuel development. The Carter administration is pledged to promoting alternative-fuel technologies, and the small synthetic-fuel sector, mostly the corporate progeny of large petroleum conglomerates, actively solicits such support.

The reasons for the private sector's interest in federal funding are clear. Most corporations entering the synthetic-fuel sector are small and have difficulty attracting the large capital investments customarily required to produce even a pilot or demonstration facility. Coal-liquefaction or -gasification plants, at these early stages, may cost between $250 million and $500 million and entail great technical and commercial uncertainties. Thus federal participation seems consistent with the national interest in energy production and the economics of the private sector.

The risks of federal involvement arise not merely from the direct ecological impacts of these fuel technologies or from Washington's determination to support their development. An equal concern is the logic of political management that imposes an overlay of calculations to spending decisions highly sympathetic to the more environmentally hazardous technologies.

**Billion-Dollar Appetites: The Politics of Energy R&D**

Portsmouth's prayers for nuclear deliverance had yet to ascend over Ohio when the DOE opened another window on the politics of R&D spending. An early 1978 conference had been arranged to announce that DOE proposed to spend between $20 billion and $40 billion through 1990 to develop synthetic-fuel systems. Among the reasons cited to reporters by a DOE official, "It is quite evident that Congress hungers for this sort of thing."[16]

Any domain of federal resource allocation should be understood, in part, as a political marketplace. Political exchanges among the actors in the allocation structure take several important forms: coalition building between elective officials and private interests, log rolling and vote trading among legislators and the White House. Goods–in this case dollars and projects–are normally and naturally exchanged by a logic as much political as scientific or economic. Questions of political advantage can never be far removed from deliberations on R&D spending, even as officials ponder the substantive merits of particular items.

Nothing in this is new. The structure and logic by which federal resources are created and exchanged within government have been exhaustively studied, as, for instance, in studies of subgovernments and congressional norms. More important is that the logic of public choice favors high energy production, centralized energy-supply systems, and synthetic-fuel systems. It is a logic rooted deeply in the institutional design of Congress and the presidency.

*Projects as Political Capital*

The U.S. electoral system, with its emphasis upon geographical representation and its congeniality to pressure-group activism, powerfully colors both congressional and presidential evaluations of a spending proposal. Congressmen sense that their political ambitions, and certainly their electoral fortunes, are tied to their ability to deliver federal goods to their home districts. Presidents, never far removed from a national election, sense in expenditures a political good to be utilized for coalition building. This is bound to affect both presidential and congressional approaches to energy R&D in predictable ways.

First, it favors large spending and big projects–those with high political impact and disguised costs where possible. "There is a bias toward politics with immediate, highly visible benefits–the myopic politics for myopic voters," writes Tufte about federal economic policy generally. "Special interests induce coalition-building politicians to impose small costs on the many to achieve large benefits for the few."[17] The psychology of high-impact, highly visible projects is a psychology of nuclear and fossil-fuel expenditures. Commercial-size coal-gasification and - liquefaction facilities, or shale-oil conversion facilities, are conservatively estimated to cost between $1 billion and $1.5 billion each. One DOE expert has calculated that the smallest commercial liquefaction facility would require about 725 employees for its operating staff and perhaps 2,500 during construction phases lasting three to four years. These facilities are likely to be perceived in Congress and the White House as economic stimulants to local economies. There is a traditional bias of this sort in federal R&D spending. Public officials, notes John Tilton in his study of the economics in federal R&D, favor "big, glamorous projects [which] attract attention and focus publicty on their sponsors."[18] Within the scientific sector, the view is widely held, and presumably promoted, that high-technology R&D is economic therapy. Notes the AAAS:

> High technology, R&D intensive industries have made an especially strong contribution to the economy. A . . . recent study is reported to have found that over the past three decades these industries have on the average increased output, productivity, and employment several times as fast as low-technology industries, while raising prices far less.[19]

The appeal to this style of political economy can be appreciated by considering the relative attractions of centralized solar and synthetic-fuel installations to federal officeholders. It has been estimated to take ten to fifteen years before most centralized solar technologies now under development will reach commercial viability sufficient to affect the national economy. The ultimate payoff from development is deferred by at least a decade for any president or legislator now supporting such systems. The only short-term payoff to be had lies in a technology's capacity to generate jobs, capital investment for developers, or economic growth where research or development is sited. Comparatively synthetic-fuel technologies generate far more visible, short-term impacts, that temporally lie close enough to the next few elections to seem relevant to the electoral fortune of a politician.

A second consequence in treating energy R&D as political exchange is a tendency to multiply the number and geographical dispersion of projects. This implies an increase in both the magnitude and duration of the environmental risks to the extent that high-technology systems are part of this trend. The pressure in these directions is generated by a legislative desire to share broadly in the delivery of these goods and the temptation of presidents to appease this interest, and inflationary strategy that mints projects to satisfy purchasing pressure. This tendency is especially common in federal public-works programs. John Ferejohn's recent study of congressional public-works expenditures noted that "the principal institutional features leading to overspending on public works are those that constitute the very basis of representative government in the United States: geographic representation, majority rule and the committee system."[20]

Federal R&D programs were exposed to this pressure early and successfully. Donald K. Price notes that the massive federal involvement in R&D programs following World War II created "all kinds of patronage" for the academic and scientific community. "Political pressure produced automatic formulas to distribute widely the early funding for agriculture and public works research."[21] When other project grants accumulated among a few universities, the others complained; regional associations and political spokesmen had to be placated. "To accommodate the various states and regions which had no universities qualified to compete for project grants," observes Price, "programs of financial support had to be invented to create new centers of excellence, able to compete on an equal basis."[22]

### *Creating Economic Dependencies*

Federal R&D expenditures, like federal transfer payments and public works, are catalysts to the formation of economically dependent interest coalitions. Indeed such federal expenditures often create new interest groups, specifically organized to promote continuing federal funding of such programs. The result is to impose rather predictable constraints on future program alterations.

First, major changes have to be negotiated on a political as well as substantive basis. Questions of program priorities, funding levels, and duration are settled, in large part, with as much concern for who is affected as for what is done. Second, dependent constituencies are quite often budget expansionists and policy conservatives; they can more easily agree upon continuing existing programs at increased funding rather than they can on substantively important policy changes. Third, incumbent politicians are tempted to manipulate funding for electoral advantage and thereby to transform dependent constituencies into dependable electoral coalitions. Finally, federal, state, and local governmental agencies with major program shares also become program partisans.

The U.S. experience with nuclear R&D illustrates these tendencies. During its expansionist phase, the program tied vast economic sectors to program spending and made them partisan enclaves for nuclear development. In the early 1970s, the AEC had contracts with 538 corporations and 223 colleges and universities. Contractors directly employed 125,000 workers in every state. Union Carbide, the largest of these contractors, had received in excess of $330 million in AEC funds by the mid-1970s. The community of interest between the AEC, the Congressional Joint Committee on Atomic Energy, and the private nuclear sector has been amply described. It is an illustration of a subgovernmental structure's exercising nearly autonomous authority over nuclear-power issues while relentlessly promoting nuclear technology. In the end, remarked present Nuclear Regulatory Commissioner Peter Bradford, the government "gravely overestimated" the promise of nuclear energy. "As bureaucratic and institutional prestige become committed, one almost rational step at a time, to stated and perceived goals that had little to do with the real national interests, truth and other people's money were the first casualties."[23]

The tendency of incumbents to use nuclear R&D for electoral advantage is suggested in figure 5-1, which charts the incremental changes in nuclear R&D funding between 1970 and 1980. Two patterns are evident: funding tends to increase, often quite sharply, during the fiscal years overlapping congressional and presidential elections, and peaks in funding increments correspond to presidential election years. Clearly an electoral calculus has affected incremental funding decisions; program expansion is a strategy intended to influence electoral outcomes. As decisions about support levels for the program are made on the basis of political advantage rather than substantive merit, powerful incentives are liberated to keep the program afloat.

President Carter's current struggle to terminate the Clinch River project (CRP) displays the political factors involved in nuclear R&D with fulsome detail. The project, intended in the early 1970s to create an experimental liquid metal, fast-breeder nuclear reactor (LFMBR) near Oak Ridge, Tennessee, had originally been estimated to cost $667 million. By 1978 the federal government had spent more than $470 million on the enterprise, now estimated to cost $2.2 billion. Washington was dispersing $12.5 million monthly to 350 individuals directly employed at the site and 5,000 others in 23 states working for the project's 124 contractors. In addition to federal administrative agencies and congressional committees with program responsibilities, about 741 private and municipal power producers shared in the project's underwriting. Early in 1977 President Carter, citing the growing danger from international proliferation of nuclear materials as a major justification, requested only $200 million for the CRP and announced his intention to terminate the project. Nonetheless, the CRP continues to be heavily funded, and the president's battle with supporters of the project has dragged through two inconclusive years. The coalition protecting the project, ranging from congressional committees across the full range of state, local, and private interests sharing in program funding, remains strong. President Carter, in fact, has discovered that he will probably be unable to stop the fast-breeder reactor program at all. In late 1978 the administration agreed to perform detailed design studies for a breeder facility two to three times the size of the CRP, apparently as a bargaining counter to weaken opposition to closing the CRP itself. Congress, however, adjourned in late 1978 after passing public law 95-482, which provided $272.8 million for the base LMFBR program, $15 million more for technology integration and $175.5 million for the CRP or an alternative approved by Congress. The

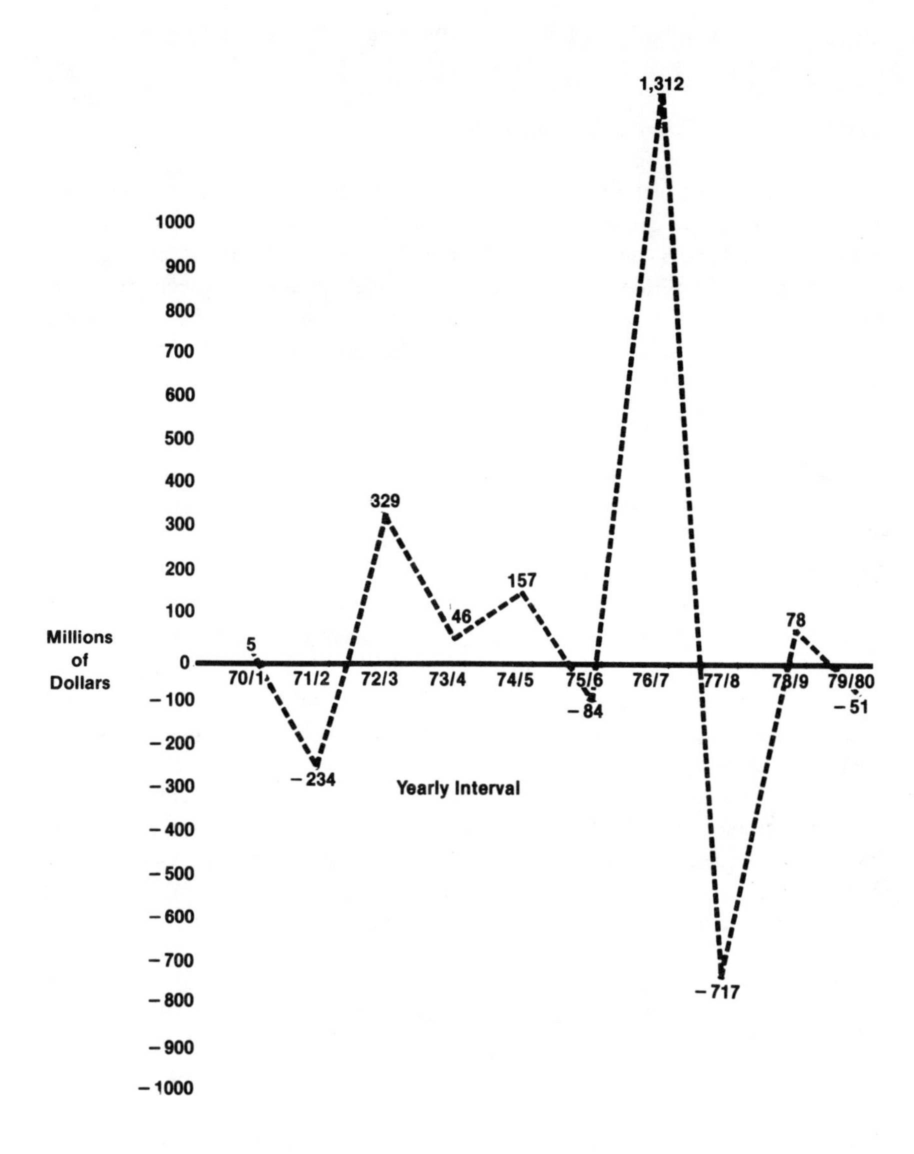


**Figure 5-1.** Incremental Changes in Budget Allocation for Nuclear Research and Development, 1970-1980

last state of the president seemed worse than the first, for he ended—temporarily, perhaps—with both the CRP and a continuing fast-breeder program.

*Leverage Points: Discretionary Authority*

The current fossil-fuel program permits the DOE considerable discretionary authority in project recommendations, budget requests, project selection, and management. This is inevitable in any R&D procedure. Congress must always delegate some discretion to administrative specialists in making complex technical decisions. Also discretionary judgments increase as projects move away from basic research and toward commercialization when issues of economic, institutional, and political feasibility increase. The significance of discretionary judgments is that they become leverage points where partisans of various programs exert influence, often successfully. Discretionary authority can often legitimate, or camouflage, an agency's decisions to bend its program to the shape of a particular public or private interest. From an ecological perspective, these discretionary judgments bear careful scrutiny because they set the environmental direction of the fossil-fuel program and are amenable to influence.

Within DOE's fossil-fuel program, several of these discretionary points merit special consideration. The first concerns projections of future energy demand. The econometric models often used to estimate the economic feasibility of energy programs—and particularly to produce the benefit-cost estimates—are highly sensitive to assumptions about future energy demand. Federal energy agencies and their consultants have a well-documented bias toward high-demand projections that rationalize federal investment in energy-supply technologies with their major ecological costs. An instructive illustration of the consequences in changing these demand scenarios is provided by the CRP. The Nixon administration estimated in 1974 that the project would yield $19.4 billion in gross benefits, all attributable to increased energy production. In 1978 the Carter administration recalculated the benefit-cost ratio for the CRP, using lower and more-plausible estimates of future power demand, and found there would be no economic benefits.[24]

The second point is selecting among competing processes for demonstration. The DOE apparently lacks a clear set of priorities and explicit criteria for choosing the energy technologies that it will recommend for development. In mid-1978, the GAO reported, "Several officials said they lack formal guidance and/or criteria for establishing synthetic fuel technology priorities."[25] Further, the factors to be considered in choosing among alternative proposals for a technology's development were "subjectively developed and considered," but officials "did not document how, or the extent to which, these factors were considered in establishing priorities."[26] Such a policy vacuum is an invitation to lobbying by proponents of various technologies, including congressmen, and can easily produce decisions based less on technical or environmental merit than upon political expendiency.

Finally is withholding federal support. The Carter administration is ostensibly committed to restricting its R&D investments more severely than prior administrations were. The Office of Management and Budget has recently announced that Washington will avoid investments in technologies "where user demand, economic viability and institutional acceptance are unlikely."[27] These constraints seem more apparent than real, however. It is often difficult to estimate in advance of a technology's extensive development whether it will prove economical or institutionally acceptable. User demand may depend upon issues not directly related to a technology's development; it may depend upon resolving issues only through development itself. Thus these spending restraints turn out to be rather soft and highly discretionary in their definition and application.

The existence of so much discretionary authority within DOE underscores the extent to which the fossil-fuel program remains a policy no man's land where substantive issues are highly politicized and their resolution dependent on the play of political forces that lack a clear, consistent commitment to environmental protection. Currently the program is moving resolutely toward future energy-production scenarios that have significant environmental risks. If history repeats itself, the political structures organized about the nuclear R&D program, with all their implicit ecological hazards, are likely to reappear within the fossil-fuel domain.

## Restoring Environmental Balances

Much can be done immediately to impose a greater regard for environmental equities in the nation's energy R&D program. Most of these proposals have been identified and advocated already. They will achieve greater urgency as the ecological implications of the present fossil-fuel program are better understood.

First, *there is a need to strengthen the review of environmental impact statements prepared within DOE for its own projects.* These statements, required by the National Environmental Policy Act of 1969 (NEPA), are intended to cause DOE and other federal agencies contemplating actions "significantly affecting the quality of the human environment" to carefully weigh the environmental impact of their programs and, if appropriate, to modify them to assure greater environmental protection. The Department's resources for NEPA review of energy R&D proposals, however, have been relatively meager from the agency's inception. The House Subcommittee on the Environment and the Atmosphere reported in mid-1978 that "funding and personnel figures appear reflective of the low importance perceived for the NEPA activities within the Department's priority setting."[28] The subcommittee found the department's interest in NEPA review so feeble in its program offices that "mechanisms appear to depend primarily upon the goodwill of each program office in contacting the NEPA Office."[29] At a minimum, reforms should include a much larger staff of NEPA specialists within the department's review office, a larger support budget for NEPA review, and a compulsory NEPA review of all R&D project grants well in advance of funding commitments for all those with significant environmental impacts—assurance, in effect, that most fossil-fuel projects will be reviewed.

Second, all benefit-cost calculations for fossil-fuel technologies should include sensitivity analysis to a broad range of energy-demand projections. An advantage of this approach is that it makes explicit the variety of different energy-demand forecasts that might be used in benefit-cost calculations, forces a consideration of the justification in selecting a specific projection, and suggests the degree to which the economic attractiveness of a particular technology is tied to a specific energy demand forecast.

Third, federal institutions responsible for establishing energy R&D policy should consider a requirement that future benefit-cost calculations, or NEPA statements, for energy projects include an estimate of the net energy production from each project and an identification of the environmental risks, or other effects, transferred from the present to immediately succeeding generations by a technology's full commercialization. In simplest terms, a net energy calculation would constitute a statement of a project's energy productivity after the energy invested in the building and operation of the energy facility is deducted from its gross energy output. Describing the environmental risks transferred across generations by a project deals with what John V. Krutilla has called "inter-generational equity." "As our means of transferring costs from the present to the future increase—for example, by generating long-lived pollutants—the need to bring inter-generational fairness into the decision-making process becomes more compelling," he argues.[30] While an obvious example of this generational transfer of risk in energy R&D arises from nuclear technology—when, for instance, problems of nuclear waste disposal are left to the future—the same logic applies to toxic substances generated by synthetic-fuel systems. Especially in light of the unknown environmental impacts to be anticipated for toxics themselves unfamiliar, the relevance of this generational balancing seems apparent. Neither procedure has been widely studied within the federal government; clearly both need greater definition and operational exploration. Both strategies, however, would force the environmental consequences to synthetic-fuel development to greater visibility publicly, bureaucratically, congressionally, and even presidentially. This would certainly be an improvement over the present status of the fossil-fuel program.

These proposals are likely to affect current patterns of federal R&D funding in at least two respects. First, the rate at which incremental spending for new fossil-fuel technologies will rise in the 1980s probably will be diminished by requiring more-rigorous economic and environmental evaluation of such projects. These evaluations are likely to force into public visibility the environmental risks and economic uncertainties attending synthetic-fuel technologies much more dramatically than would otherwise be the case. It will be considerably easier to alert and to mobilize interests advocating environmental protection and economic prudence in public investments, a coalition not usually found working the same side of public issues. This mobilization probably will force greater public debate on fossil-fuel technologies and thereby protract their rate of evolution. This will certainly not stop, or reverse, the trend toward greater federal investment in fossil-fuel R&D, but it will be a constraint upon an extremely rapid federal development of the synthetic-fuel sector propelled by excessive public alarm at prospective energy scarcity. A more protracted rate of funding will provide more time for the comparative attractions of solar technologies, energy conservation, and other environmentally benign energy strategies to become apparent.

These proposals would also encourage a greater caution toward synthetic-fuel technologies among that amalgam of private interests currently advocating greater constraint upon public spending. The economic luster to most fossil-fuel technologies now contemplated is likely to dim appreciably when they are subjected to close economic scrutiny. This fact is not yet widely advertised in much public discussion of the fossil-fuel technologies. However, the movement to control public spending could well find in the federal fossil-fuel R&D philosophy much to oppose. One result of this opposition might be to force upon the private sector a greater responsibility for finding investment capital to create fossil-fuel-production facilities. This might create not only greater caution in the development of these technologies but might considerably improve the speed with which acceptable ones are developed. In any event, the shaky economic underpinnings of the proposed new fossil-fuel technologies are likely to work against current federal approaches to energy R&D when the weakness becomes public knowledge. All this, in turn, will diminish the environmental impacts of a new raid on coal.

The need to create countervailing forces to rapid synthetic-fuel development is especially important as the United States enters the 1980s and the Carter administration advocates a number of measures, such as the president's energy mobilization board, intended to speed the construction of new energy-production facilities. This board is meant to speed the procedural review of new energy facilities and, in the process, to diminish the constraining force of environmental evaluation required in these procedures. It would be most unfortunate for the United States to enter the 1980s—only a decade away from the first Earth Day—committed simultaneously to minimizing the environmental safeguards while maximizing the prospects for ecologically dangerous energy technologies.

## Notes

1. The Portsmouth incident is discussed in the *New York Times,* July 7, 1977.

2. Statistics on federal R&D expenditures since 1970 are conveniently summarized in National Science Foundation, *Federal Funds for Research, Development and Other Scientific Activities, 1977* (Washington, D.C.: Government Printing Office, 1978), and, for later years, Executive Office of the President, Office of Management and Budget, *Special Analysis: Budget of the U.S. Government, 1978, 1979, 1980* (Washington, D.C.: Government Printing Office). Hereafter these will be cited as "NSF Analysis" and "OMB Special Analysis."

3. Congressional Budget Office, *Energy Research, Development and Demonstration* (Washington, D.C.: Government Printing Office, 1977), p. xiv.

4. Willis H. Shapley and Don I. Phillips, *Research and Development: AAAS Report III* (Washington, D.C.: American Association for the Advancement of Science, 1978), p. 20.

5. Paul F. Rothberg, "Coal Gasification and Liquefaction," Congressional Research Service, Issue Brief IB77105 (Washington, D.C.: Congressional Research Service, 1978). Hereafter cited as "CRS Coal Study."

6. Edward R. Tufte, *Political Control of the Economy* (Princeton, N.J.: Princeton University Press, 1978), p. 139.

7. Rodney W. Nichols, "Some Practical Problems of Scientist-Advisors," *Minerva* (October 1972): 603-613.

8. Consumption estimates are based on figures contained in U.S. Department of Energy, *Environmental Readiness Document: Coal Gasification* (Washington, D.C., September 1978), p. 9, and *Environmental Readiness Document: Coal Liquefaction* (Washington, D.C., September 1978), p. D-2. Hereafter cited, respectively, as "DOE Gasification Study" and "DOE Liquefaction Study."

9. CRS Coal Study, p. 13.

10. DOE Gasification Study, pp. 2, 12, 14.

11. DOE Liquefaction Study, p. 12.

12. Paul F. Rothberg, "Oil Shale Development: Outlook, Current Activities and Constraints," Congressional Research Service, Issue Brief IB74060 (Washington, D.C.: Congressional Research Service, November, 1978), p. 2.

13. OMB Special Analysis, 1979, sec. P.

14. Cited by Richard S. Lewis, *The Nuclear Power Rebellion* (New York: Viking Press, 1972), p. 28.

15. General Accounting Office, "Fossil Energy Research, Development and Demonstration: Opportunities for Change," Report EMD 78-57 (September 18, 1978), p. 31. Hereafter cited as "GAO Fossil Study."

16. *New York Times,* February 25, 1978.

17. Tufte, *Political Control,* p. 143.

18. John E. Tilton, "The Public Role in Energy Research and Development," in Robert J. Kalter and William A. Vogley, eds., *Energy Supply and Government Policy* (Ithaca, N.Y.: Cornell University Press, 1976), p. 115.

19. Shapley and Phillips, *Research and Development,* p. 78.

20. John A. Ferejohn, *Pork Barrel Politics* (Palo Alto, Calif.: Stanford University Press, 1974), p. 252.

21. Donald K. Price, "Money and Influence: The Links of Science to Public Policy," *Daedalus* (Summer 1974): 101-102.

22. Ibid.

23. *New York Times,* July 9, 1978.

24. Marcia S. Smith, "Breeder Reactors: The Clinch River Project," Congressional Research Service, Issue Brief IB77088 (Washington, D.C.: Congressional Research Service, December 1978), p. 9.

25. GAO Fossil Study, p. 20.

26. Ibid.

27. OMB Special Analysis, 1980, sec. P.

28. U.S. Congress, House, Committee on Science and Technology, Subcommittee on the Environment and the Atmosphere, *Oversight: Environmental Responsibility Within the Department of Energy,* Document 108 (Washington, D.C.: Government Printing Office, 1978), p. ix.

29. Ibid.

30. John V. Krutilla and R. Talbot Page, "Energy Policy from An Environmental Perspective," in Kalter and Vogley, *Energy Supply,* p. 82.

# 6 Energy-Environmental Trade-Offs in the Courts: Nuclear and Fossil Fuels

*Lettie McSpadden Wenner*

The 1970s, which began as the environmental decade, increasingly became defined by conflicts over energy development. The courts, like other branches of government, were called upon to decide such issues as which forms of energy are more economical and less damaging to the public health and physical environment than others. To an even greater degree than legislators and executive officials, judges were able to consider only discrete questions about individual cases. Yet their decisions on a case-by-case basis have had an impact on the general pattern of energy policy. Overall numerous discrete decisions have cumulated with decisions made elsewhere to affect strategically the availability of different forms of energy to the American public.

Although it is generally agreed that the United States lacks a comprehensive energy policy, nevertheless laws and agencies proliferate that influence the forms and quantities of energy we use. Each time a court renders a decision about whether a dam may be built, land condemned for an electric transmission line, or supertankers admitted to a coastal port, there is some impact on the quantity and form of energy available in some area of the country. In the past courts have played an important role in interpreting such major pieces of environmental legislation as the National Environmental Policy Act[1] and the Clean Air Act.[2] In the future their role can only increase as various litigants call upon them to rule on the legality of regulations written about newer pieces of legislation such as the Surface Mining Control and Reclamation Act.[3]

Two forms of energy that are often considered to be in direct competition with each other are nuclear energy and fossil fuels. Both can be used to generate electricity. Each produces one or more environmental contaminant (radiation or sulfur oxides and particulates, respectively) whose impact on public health and welfare is difficult to measure and compare. Estimates of abatement costs for avoiding such injuries have been equally controversial, and policy makers have been unwilling or unable to make direct comparisons between these two forms of energy. The courts, therefore, have been called upon frequently to rule on the adequacy of the safeguards and controls placed upon individual plants of each type.

Two major patterns in judicial decision making have influenced the manner in which courts have decided these issues. The first is the tendency for judges to defer to the technical expertise of administrative agencies in civil actions involving the government. This tendency is exemplified most clearly by the courts' willingness to defer to the judgment of nuclear engineers in the Nuclear Regulatory Commission (NRC, formerly the Atomic Energy Commission) in cases concerning nuclear safety. Because of this agency's promotional attitude toward nuclear power, judicial deference has had the effect of encouraging the development of nuclear power. In cases involving fossil-fuel-fired plants, however, the government's technical expertise has been represented by the Environmental Protection Agency (EPA), whose expertise is directed toward controlling air and water pollution and consequently leads to slower and more restrictive development of power. Any judicial deference paid to EPA is likely to have a dampening effect on the construction of fossil-fuel-fired electric plants. The cumulative impact of the two kinds of decisions may well result in a preference for nuclear over fossil-fuel-fired generating plants, even though the courts have made no deliberate choice between the two forms of energy.

The other major pattern in judicial decision making that has affected energy-environmental trade-offs concerns the issue of federalism. Deciding who should decide ultimately may determine what substantive decision will be made. The courts have an important responsibility when they choose between the central government and the states and localities. Because of the nature of the legislation that controls nuclear power, the courts traditionally have viewed this particular form of energy as an exclusive domain of the federal government. But recently cases have arisen in which some states have argued that they should be the ones to determine the level of risk assumed by their residents. The major piece of legislation controlling fossil-fuel plants, the Clean Air Act, contains a much more positive and discretionary role for state governments. But a considerable debate has arisen over the shared responsibilities of the central government and the states for controlling pollutants in the atmosphere, and the courts have often been called upon to referee such disputes.

The two issues of administrative discretion and federalism have affected most judicial decisions about all sources and uses of energy, from hydroelectricity to gasoline-powered engines. One way to discuss the issues in a relatively limited context is to consider only court decisions having to do with the generation of electricity, first from nuclear sources and second from fossil-fuel-fired plants.

## Nuclear Cases

One of the largest categories of energy-related court cases is that involving nuclear energy. Until the Three Mile Island accident in March 1979, general public support for nuclear power was high, but throughout the 1970s a number of national public-interest groups (for example, the Natural Resources Defense Council) and locally organized groups opposed the construction and licensing of individual power plants. In addition, environmentalists opposed generic nuclear programs not specifically directed at power plants.

*Individual Plant Protests*

The earliest case occurred in January 1969 when the states of New Hampshire, Vermont, and Massachusetts combined to appeal the decision of the Atomic Energy Commission (AEC) to give the Vermont Yankee Nuclear Power Corporation permission to build a plant on the Connecticut River. The U.S. Court of Appeals for the First Circuit reluctantly held that nothing in either the Atomic Energy Act or the Federal Water Pollution Control Act forced the AEC to consider anything other than radiation hazards in deciding whether to license the construction and operation of nuclear plants.[4]

Another early case involving federalism was the Northern States Power Company case, which occurred in Minnesota in 1970. There the pollution-control agency tried to apply more-stringent radiation standards to a nuclear plant than the AEC had applied. The federal district court in Minnesota supported the power company, ruling that the AEC had preempted all regulatory authority over radioactive emissions and the state authorities were without recourse.[5] Both the Eighth Circuit and the U.S. Supreme Court were quick to affirm judgment.[6] Undaunted by this outcome, the Illinois Pollution Control Board set its own radioactive water-emissions standards for the Dresden plant in Illinois, but it was quickly struck down by an Illinois state court, citing the Northern States Power case.[7] It is clear that the preemptive powers of the federal government overcome any potential desire on the part of state government to impose stricter controls on nuclear-power plant operations.

This obvious deference in the regulatory field, however, does not necessarily preclude states from making negative decisions concerning the siting of nuclear plants within their jurisdiction. The public utility commission (PUCs) of the various states have the authority to issue certificates of convenience and necessity for the siting of such plants. For the most part, the states have proved receptive to nuclear-power plants, and the state courts, like the federal ones, have deferred to the expert judgments of the technologists in the PUCs.

This is not to say that all state decisions have gone unchallenged by opponents of nuclear power. In January 1970 the United Mine Workers, concerned about unemployment rates in the coal industry, objected to the certificate of convenience and necessity given by the Colorado PUC to the Fort St. Vrain plant, but the Colorado Supreme Court upheld the PUC's discretion to determine the economic viability of the plant.[8] In February 1978 a New York State court found that a local zoning board could not halt the building of the Indian Point nuclear plant by refusing a zoning variance for the height of a cooling tower that was a necessary addition to the plant. While the trial court argued that the NRC had preempted state authority in this matter by requiring cooling towers, the appellate court simply found that local zoning boards cannot exclude utilities when the appropriate state commission has given its consent.[9]

Some controversies over nuclear plants have involved both state and federal courts simultaneously. Regardless of the type of court, the decisions have tended to favor administrative discretion. In February 1970 a Florida property owner objected to condemnation of his land to provide passage for wastewater from the Turkey Point nuclear plant to Biscayne Bay, but both the lower state court and the Florida Supreme Court turned down his plea.[10] At the same time a federal district court in the Fifth Circuit was refusing the U.S. Department of Justice authority to control thermal emissions from Turkey Point because the Federal Water Pollution Control Act contained no provision to control them.[11]

Because of the principle of federal preemption, most cases concerning nuclear-power plants have been adjudicated in federal courts. In July 1970 a local ecology group allied with the Sierra Club challenged the South Haven, Michigan, plant on the grounds of thermal pollution. The groups took their complaint to the U.S. Courts of Appeals for the Seventh Circuit and for the D.C. Circuit, and both refused them redress on the grounds that the AEC's action was not yet final.[12] The D.C. Circuit, however, indicated that it believed the AEC would consider all relevant matters, including thermal pollution, and the court would consider an appeal if that did not prove to be the case. Similarly in April 1971 a local group asked a federal district court in New York to force the AEC to consider more than radiation hazards but were told that they should exhaust their administrative remedies before requesting court action, and then to approach the court of appeals.[13]

Later, in July 1971, the D.C. Circuit was afforded an opportunity to determine whether the AEC's actual implementation of the National Environmental Policy Act (NEPA) was as complete as the court had assumed it would be in the Michigan case. In the landmark Calvert Cliffs case, the D.C. Circuit found the AEC's procedures for NEPA compliance lacking. As the lead agency in controlling both the construction and licensing of nuclear plants for operation, the AEC was ordered to consider the cumulative impact of all environmental hazards created by one plant, not simply the radiation problem.[14]

Other local environmental groups and public-interest organizations were quick to follow the Calvert Cliffs lead and insist on the full procedural requirements by the AEC in writing environmental impact statements (EISs) for all reactors, even those already constructed before NEPA went into effect on January 1, 1970. In December 1971 the Izaak Walton League obtained a temporary injunction from the D.C. Circuit while an EIS was completed for the operating license of the Quad City plant in Illinois.[15] In 1972 the Monticello plant in Minnesota was challenged and an EIS written for it, although no injunction was granted during the process.[16] In April 1972 the D.C. Circuit applied the Calvert Cliffs precedent to the Davis-Besse plant in Ohio but sent the case back to the AEC to determine for itself whether to suspend construction while it wrote the statement.[17] Despite the fact that there is no requirement within NEPA for public hearings, nevertheless the D.C. Circuit made another innovative ruling in 1973 when it held that the Atomic Energy Act together with the Administrative Procedures Act provided sufficient support for an environmental group to insist that a public hearing be held before the AEC issued a license to the Donald C. Cook plant in Indiana.[18]

During the mid-1970s, however, the AEC and its successor, the NRC, generally complied sufficiently with the procedures of NEPA to gain judicial approval despite numerous attacks on the safety of individual plants. In June 1974 the D.C. Circuit reviewed the AEC licensing of the Pilgrim plant in Massachusetts and found that it was a technical decision that only the AEC was equipped to handle.[19] In December 1975 the D.C. Circuit reviewed the EIS for the Peach Bottom plant in Pennsylvania near Baltimore and remanded it to the NRC for reconsideration of the radioactive emissions standards it had set, which were uniform nationally and did not take the different risks of individual plants into consideration. For the major portion of the decision, however, the NRC's discretion was upheld.[20] In December 1975 the D.C. Circuit also agreed that the NRC had followed correct procedures in writing the EIS for the Maine Yankee Nuclear facility despite the argument that it should have limited the permit to fewer than forty years.[21] In March 1976 the D.C. Circuit acquiesced to a Virginia plant despite questions raised about its location on a geological fault line. It was decided that NRC's expertise in technical matters was superior to any second guessing by lay judges.[22] The Second Circuit told a group of citizens concerned about a plant in Rhode Island to exhaust their administrative remedies through NRC in April 1977.[23] A group in Kansas challenged the Wolf Creek installation because NRC did not consider a gas-fired alternative, but the D.C. Circuit turned them down in January 1979.[24]

In challenging the licensing of individual nuclear plants public-interest groups were innovative in the kinds of issues that they believed the AEC and later the NRC should consider in any EIS. Yet for the most part these issues were rejected after the initial need to have the AEC go beyond radiation hazards was established. In February 1974 Ecology Action argued before the U.S. Court of Appeals for the Second Circuit that two other issues, fuel conservation and disposal of radioactive wastes, should be considered in EISs, but the court did not agree.[25] Two years later, however, the same issues were raised in two separate cases heard by the D.C. Circuit, and both were accepted as reasons for delaying the issuance of a license. On July 21, 1976, the D.C. Circuit made two of its furthest-reaching decisions on nuclear power, indicating that it was prepared to consider not only the procedural requirements of NEPA but also the substantive content of EISs. It ordered the NRC to consider the costs of disposing of nuclear wastes in its EIS before issuing a permit for the Vermont Yankee plant,[26] and it delayed the start of a reactor in Michigan when it ruled that its EIS should consider the alternatives of conserving energy rather than simply increasing the supply.[27]

On April 3, 1978, the U.S. Supreme Court, in its severest rebuke to a circuit court about nuclear power regulation, told the D.C. Circuit to intrude no further into the regulatory process. According to the Supreme Court, the NRC alone should determine what issues are important enough to be considered in licensing hearings; if it does not wish to include conservation of energy and waste-disposal problems, it need not do so. In the words of Justice Rehnquist (to which there were no dissents):

> Nuclear power may someday be a cheap safe source of power or it may not. But Congress has made a choice to at least try nuclear energy, establishing a reasonable review process, in which courts are to play only a limited role. The fundamental political questions appropriately resolved in Congress and in the state legislatures are not subject to reexamination in the federal courts in the guise of judicial review of agency action.[28]

Laws other than NEPA have been utilized by environmental groups to delay the start-up of nuclear plants. These, too, have met with mixed results. The Sierra Club objected to an exchange of marshland by the Department of Interior to accommodate the Davis-Besse plant on Lake Erie in Ohio, but the Sixth Circuit proved unreceptive to its arguments.[29] In 1973 residents in LaSalle County, Illinois, tried to prevent a utility from acquiring land in their area, but a federal district judge ruled that no government action had yet taken place.[30] The D.C. Circuit had earlier refused to hear the case because the farmers had not addressed their grievance to the AEC.[31] Throughout 1978 New Hampshire environmentalists tried to delay construction of the Seabrook plant. They succeeded in getting the EPA to review its permit for a wastewater discharge[32] and convinced the NRC to reroute some of the transmission lines from the plant.[33]

Another group that proved tenacious in its efforts to delay the opening of a plant was the Carolina Environmental Study Group, which sued the AEC for giving a permit to the Duke Power Company in 1975.[34] Rebuffed by the D.C. Circuit, which found the AEC's impact statement sufficient, the group developed the novel argument that the Price-Anderson Act, which relieved industry of most of its liability for damages done by nuclear accidents, was unconstitutional. In 1977 a federal district judge in North Carolina agreed that the law was unconstitutional because it did not afford neighbors of nuclear plants equal protection of the law and deprived them of property without due process.[35] In July 1978 the Supreme Court overturned this decision too, with Chief Justice Burger writing for four other justices. Three others would not have reached the merits of the case, but all agreed that the district court had overstepped its role when it decided that the act was unconstitutional.[36]

Earlier two other circuits had proved more receptive to nuclear skeptics' ideas than was the Supreme Court. In February 1974 the Colorado Public Interest Group sued to have the EPA set standards for radioactive discharges to water. That issue had been settled by administrative decision in 1973 when the AEC was given total control over all nuclear-power questions. The federal district court acquiesced in the interest of bringing the Fort St. Vrain plant on line as quickly as possible.[37] However, the Tenth Circuit overturned that restrictive ruling and found that the 1972 amendments to the Federal Water Pollution Control Act, coming as they did after the Atomic Energy Act of 1954, could be read as controlling at least the radioactivity of wastewater.[38] In June 1976, however, the Supreme Court unanimously reversed and returned all authority to the NRC.[39] In April 1975 the Seventh Circuit decided that the Bailey nuclear plant being constructed in the Indiana dunes was too close to a population center and violated the NRC's own regulations.[40] This decision was reversed and remanded to the Seventh Circuit without comment from the high court.[41] And in April 1976 the Seventh Circuit reluctantly found for the AEC concerning the Bailey plant after its rebuff from the Supreme Court.[42] Whether the issue is one of compliance with NEPA or the NRC's own regulations, it seems that the U.S. Supreme Court has decreed that the federal circuits are to defer to regulatory-agency discretion in licensing individual plants.

### *Generic Nuclear Programs*

In addition to cases involving licensing of individual power plants, public-interest groups have raised other kinds of challenges to the nuclear power program. Many cases have been argued on NEPA grounds, as in 1971 when a group tried to prevent an underground nuclear test in Alaska. The D.C. Circuit overturned the district court, which had given summary judgment for the AEC, arguing that the lower court must consider the completeness of the EIS. The district court later found that it needed certain documents in order to make a decision, which the AEC refused to surrender. Although the circuit upheld the district's demand for the documents, it refused to issue an injunction to halt the test while the information was obtained.[43] In a related case, several members of Congress sued under the Freedom of Information Act for release of the same executive documents, but the Supreme Court found a blanket exception to the act for all national-security documents and those associated with them in the same file.[44]

In another dispute unrelated to power generation, the Second Circuit in July 1973 upheld the AEC's decision not to write an EIS about a research reactor in Columbia University.[45] The Supreme Court denied review, with Justice Douglas dissenting.[46] In December 1978 a federal district court overruled any attempt by New York City's health authorities to control the reactor, saying that the federal controls preempted any local concerns.[47] In March 1970 a public-interest group in Colorado tried to prevent the AEC's experiment with atomic explosions in order to release gas trapped beneath the surface.[48] The district court in Colorado believed that the AEC had the necessary expertise to determine whether the experiment was safe.

Other NEPA cases proved more successful for environmental groups. In August 1974 the Sierra Club won a case to force the AEC to write an EIS about its exports of nuclear materials.[49] In June 1973 the D.C. Circuit decided that an EIS must be written for the breeder reactor program as a whole rather than simply parceling out decisions for each plant.[50] But this decision backfired for another group in Tennessee when it challenged the Clinch River breeder reactor in June 1975. A district court in the District of Columbia this time determined that the programmatic EIS mandated by the circuit was sufficient to cover all aspects of the project. Simply because the Energy Resource and Development Agency had asked for additional funds did not mean that a new EIS must be written.[51]

In July 1973 Ralph Nader and the Friends of the Earth attempted to get a federal district court in the District of Columbia to rule on AEC's regulations for the safe operation of twenty different nuclear plants, but the case was sent to the circuit.[52] There, in May 1975, the circuit most favorable to environmentalists found that the NRC had the requisite expertise to make decisions about all twenty plants and rejected the plea that they were unsafe.[53] In January 1979 an individual took an even broader approach and argued that all nuclear plants should be closed on health grounds. Both the Sixth Circuit and the D.C. Circuit dismissed this claim, citing the Supreme Court's Vermont Yankee decision and arguing they had no jurisdiction.[54]

In May 1976 the Natural Resources Defense Council (NRDC), one of the most active antinuclear litigants, challenged the safety of the nuclear industry's program to reprocess spent fuel by using plutonium and uranium in a new type of reactor. The U.S. Court of Appeals for the Second Circuit was sympathetic to the argument that an EIS was needed for the entire program before any individual license for a plant using the mixed fuel could be issued.[55] But the U.S. Supreme Court in 1978 vacated that judgment without explanation.[56]

By the late 1970s the waste-disposal issue had become as important as plant licensing had been earlier. In February 1977 the state of New York tried to prevent any radioactive wastes from being shipped in or out of New York airports, but the Second Circuit refused to issue an injunction.[57] In May 1978 the storage of nuclear wastes in Oregon was challenged by environmentalists who argued that "temporary" storage of wastes constituted permanent storage, since there was no adequate long-term solution. The federal district court there, however, found that storage was temporary regardless of the length of time involved, since the federal authorities were working on a longer term solution.[58] In May 1978 the NRDC attacked the practice of storing wastes at reactor sites on the grounds that they should be licensed as well as studied by the NRC. Earlier, in 1973, the AEC had agreed to write a programmatic EIS for waste storage, and the D.C. Circuit agreed that individual EISs were needed at temporary storage sites. However, the court disagreed with NRDC that NRC licenses should be issued before wastes could be temporarily stored near reactors.[59] Two months later, in July 1978, NRDC went on to argue against NRC's licensing additional reactors until a way was found to dispose of nuclear wastes permanently. However, the Second Circuit, reacting to the Supreme Court's earlier lecture to the D.C. Circuit about interfering with the regulatory process,[60] turned down this argument.[61] In a separate case in January 1979, the state of Illinois challenged the NRC's wisdom in maintaining a radioactive disposal site in Morris, Illinois, but the U.S. Court of Appeals for the Seventh Circuit left such decisions to the NRC, also quoting the rebuke from the high court.[62]

### *Summary of Nuclear Cases*

Some fifty-eight court cases, representing some forty-six controversies, were decided on nuclear questions in state and federal courts below the U.S. Supreme Court level in the 1970s. Of these, only thirteen were decided in favor of challengers to the industry, and four ended in compromise. (See table 6-1.) Of the thirteen victories for environmentalists, seven were later reversed by the U.S. Supreme Court. Of even greater interest than the poor showing by environmental interests in the ratio of wins and losses is the fact that nearly all of the cases were initiated by antinuclear groups. The extreme deference shown by courts at both the federal and state levels has not been due to severe treatment of the industry by the regulatory agency. Of fifty-four cases only four were started because of industry's complaints against stringent regulation. The other fifty cases were due to the belief of various individuals and groups that the regulations were too lax, including the six initiated by government. Most of these were state government actions challenging the federal preemption of control over nuclear plants.

Numbers of demands made on the various circuits of the federal system have varied widely. (See table 6-2.) Clearly environmentalists felt that their best hope lay with the District of Columbia Circuit, which has proved most receptive to environmental actions in the past. Groups from all over the country that are hopeful of delaying or improving the safety characteristics of their local nuclear generators have appealed to this circuit, and sometimes

**Table 6-1**
**Court Decisions, by Initiators of Cases**

| | *Fossil-Fuel Plants* | | *Nuclear Plants* | |
|---|---|---|---|---|
| *Initiator* | *Pro-Power* | *Pro-Environment* | *Pro-Power* | *Pro-Environment* |
| Environmentalists | 16 | 9[a] | 32 | 12[c] |
| Government | 8 | 6[b] | 6 | |
| Industry | 13[b] | 17 | 3 | 1 |
| Total[d] | 38 | 31 | 41 | 13 |

[a]Includes two cases later reversed by the U.S. Supreme Court

[b]Includes one case later reversed by the U.S. Supreme Court

[c]Includes seven cases later reversed by the U.S. Supreme Court

[d]This table does not include cases decided in a neutral or ambiguous manner.

successfully. In addition, the D.C. Circuit has ruled on most issues of national importance such as the breeder reactor and the waste-disposal problem. Of some twenty-four cases heard in the D.C. Circuit at both trial and appellate levels from 1970 to 1979, ten were decided in favor of the nuclear regulators, three were ambiguous, and eight were decided in favor of the nuclear critics. Of these eight, three were later overturned by the U.S. Supreme Court. In the Second Circuit, with the second largest number of nuclear cases, one case was

**Table 6-2**
**Ranking of Circuits, by Decision Pro-Environment**

| | *Fossil-Fuel Plants* | | *Nuclear Plants* | |
|---|---|---|---|---|
| *Circuit* | *Percent Pro-Environment* | *Number of Cases* | *Percent Pro-Environment* | *Number of Cases* |
| First | 100 | 2 | 33 | 3 |
| Sixth | 58 | 12 | | 2 |
| D.C. | 56 | 9 | 33 | 24 |
| Second | 50 | 6 | 10 | 10 |
| Fifth | 50 | 6 | | 3 |
| Eighth | 50 | 4 | | 3 |
| Third | 42 | 12 | | 0 |
| Seventh | 40 | 10 | 17 | 6 |
| Ninth | 33 | 6 | | 2 |
| Fourth | | 2 | 100 | 1 |
| Tenth | | 4 | 25 | 4 |
| Total | | 71 | | 58 |

Note: the total number of cases varies from table 6-1 to table 6-2 because ambiguous cases were not included in table 6–1.

decided in favor of environmentalists (and overturned by the U.S. Supreme Court) and nine in favor of government. Of the six cases that the Seventh Circuit heard, one was decided in favor of opponents to a power plant (and overturned by the high court). The Tenth Circuit decided one of four against the government, and this too was overturned. The First Circuit decided one of three cases in favor of the protestors, a minor procedural decision that was soon resolved by the NRC. The Eighth Circuit decided none of its three cases in that direction, although one was an ambiguous decision. The Fifth Circuit resolved all three of its cases in the nuclear industry's favor; the Sixth and Ninth decided both of their cases in the same manner. The Fourth Circuit heard only one case, in which the Price-Anderson Act was declared unconstitutional, but the Supreme Court quickly reversed. To date, no decision favorable to the nuclear industry has been overturned by the Supreme Court.

For the most part, NEPA and non-NEPA cases involving the licensing of individual plants, as well as programmatic nuclear cases, have revolved around one major question in administrative law: the degree to which courts may oversee the administrative discretion of agencies created to regulate a highly technical process. The courts began their review rather timidly, without much statutory law to support judicial oversight. After the passage of the NEPA, however, some of the circuits, most notably the District of Columbia, began applying strict procedural requirements to the AEC and later to the NRC. After a period of quiet in which it appeared that the NRC had acclimated itself to the procedural requirements, the D.C. Circuit again became active, this time substantively reviewing the completeness of EISs. Other circuits, too–the Second, First, Fourth, Tenth, and Seventh–showed some willingness to restrain the NRC in what the latter obviously felt was its own field of expertise, and the commission turned to the Supreme Court to redress the balance of power. The Burger Court has proved especially willing to hear nuclear-power appeals that have been decided in favor of nuclear critics and overturn them, warning the circuits against interfering with the administrative discretion of the NRC.

The issue of federalism has often appeared in combination with the balance of power between judicial and administrative decision makers; in other cases it constituted the only legal question. Although less obvious than the administrative-discretion problem, the question about whether federal regulation should preempt the states' police power to protect the health and welfare of their residents may prove to be of equal importance soon. In these cases the federal courts have generally supported the AEC or NRC just as they have in the administrative cases, despite the tendency of the U.S. Supreme Court in recent years to adopt a more-favorable stance toward state powers in other areas. It may be that in the future these two goals of the court, to defer to nuclear experts and to decentralize power, will come into conflict, but this has not yet happened.

The majority of both types of cases, those involving judicial oversight of administrative decisions and federalism cases, have been decided in a manner generally favorable to the nuclear industry. No nuclear plant has been prevented from becoming operational. Even cases such as Calvert Cliffs, which represents the most famous victories for environmentalists, ultimately resulted in the opening of the nuclear plant in question. Some cases may have been instrumental in causing the AEC or NRC to impose stricter standards on the plants and in delaying the opening of plants, in some cases for several years. Such delays have doubtless contributed to the recent reduction in the number of new nuclear plants because of escalating costs, some of which must be attributed to the imposition of stricter safety standards. Nevertheless, the courts have not proved to be a very effective deterrent to nuclear power in the 1970s.

### *Future Directions*

The rapid expansion of environmental- and consumer-protection legislation in recent years has transferred many conflicts from private tort law to the public-law sector. This development has been affirmatively sought by many public-interest law groups because private parties seeking compensation for injury after the fact have faced an uphill battle because of the difficulty of demonstrating a direct causal link between an individual polluter's behavior, an environmental hazard, and an individual victim's injury. Traditionally such cases have been unsuccessful because of the tremendous burden of proof the plaintiff faced, together with the lack of any effective legal remedy even in those cases that were won. The only long-term solutions occurred in cases where damages assessed against discharges were so high that they were deterred from repeating the same kind of accidental or negligent exposure of the public to the same hazard.[63]

Theoretically public-law remedies seem to be more promising than private damage suits. Such laws and regulations enlist the authority of government on the side of the potential victims and emphasize prevention of injury rather than compensation after the fact. All of the cases reviewed thus far have fallen into the category of public-law cases, however, and nearly all were initiated by groups dissatisfied with the implementation of regulations by the responsible agencies. Law courts are often viewed as the last resort of groups unhappy with the public policy outputs of other institutions, legislatures and administrative agencies. In nuclear-power cases courts have afforded another procedural step but not much substantive change in the policy already promulgated because of the twin tendencies for courts to defer both to the federal authority and to expert decisions in administrative agencies.

Dissatisfaction with the regulation of the nuclear industry by public law may lead to a resurgence in private damage suits. This kind of case has never been abandoned and is always potentially available to anyone who feels unprotected by public authorities. One such case was tried in connection with nuclear power in September 1972 when a group of residents of Utah sued the AEC because they believed they had been injured by fallout from the Nevada atomic tests conducted in the 1950s. Because of the difficulty in pinpointing one cause of any particular disease that becomes evident only years after exposure to radiation, the court turned them down.[64] Although there has been considerable evidence concerning radiation, cancer experts still disagree about the nature of that linkage. Many more cases of this type have been recently filed and the trend is likely to continue.

Incidents in the distant past, whose symptoms may be manifesting themselves today, and more recent events are both likely to generate such cases. The Three Mile Island accident, the most publicized of such incidents, may lead to considerable litigation in the private law field. One case already discussed under public law, the North Carolina challenge to the Price-Anderson Act, has already raised the issue of liability for private damage.[65] One reason for the Supreme Court's rejection of the environmentalists' argument was the belief that no one had yet been hurt. Given the demonstrable economic loss of employment and commerce in the Harrisburg area, the costs of evacuation by those forced to leave Middleton, Pennsylvania, as well as the psychological damage done to individuals, it seems to be an opportune moment to test the willingness of state courts to impose strict liability on the nuclear industry. Proponents of Price-Anderson argue that the law's statutory liability limit and federal governmental assumption of responsibility are better safeguards for individual victims of nuclear accidents than are varied state tort liability standards. It appears that we now have a chance to test some of those theoretical assertions in a less-than-catastrophic setting. Whatever the outcome of individual cases, however, it is safe to assume that private cases of this type will proliferate, given the Supreme Court's extreme deference to the NRC in public-law cases.

## Fossil-Fuel Plants

The fossil-fuel-fired plant is another type of energy-conversion facility that has been affected by environmental legislation passed in the 1970s. There are, however, two major differences between the regulation of such plants and the regulation of nuclear plants. First, the agency that has primary responsibility for controlling the environmental impacts of fossil-fuel-fired plants is the EPA, which was created in 1970 to reduce environmental degradation, not to promote any particular form of energy. Second, the Clean Air Act (CAA) is the legislation under which most restrictions are placed on fossil-fuel plants. The CAA, unlike the Atomic Energy Act and the NEPA, is a minutely detailed law, often described as innovative or "technology forcing." The complex language of the act has led to considerable judicial interpretation, but the specific requirements in the CAA have given courts less discretion than have NEPA and AEA.

Because of the nature of the enforcement agency and the relevant law, litigation about air-pollution controls for power plants has assumed a much different tone from that surrounding nuclear-power plants. Court cases about nuclear power have been introduced nearly exclusively by nuclear skeptics, challenging the discretion of government nuclear regulators, who have served primarily as representatives of the nuclear industry's point of view. A modest number of these kinds of cases have been filed against fossil-fuel-fired plants. Organizations such as the Sierra Club and the Natural Resources Defense Council (NRDC) have been instrumental in bringing suits arguing that the EPA has not controlled the industry sufficiently closely.

Two other types of court cases not often encountered in the nuclear-power discussion are more common for fossil-fuel plants. One type is brought by government against industry for not complying with the regulations imposed by the EPA or state pollution-control agencies. Nuclear-power cases initiated by government are primarily conflicts between federal and state regulators. This indicates a different kind of relationship between nuclear plant operators and their regulators from that between electric utilities and the EPA. Although we often hear of nuclear plants being fined by the NRC, there are no cases recorded in which NRC had to take the industry to court in order to obtain compliance with its orders. Rather the NRC's normal role is to protect the industry against efforts by state agencies to control it more closely. In the fossil-fuel-fired industry, the experience has been much different. There, industry has taken an obdurate stand against the policy developed in EPA that plants should clean their emissions, preferably with flue gas desulfurization units (scrubbers), rather than with intermittent controls, tall stack dispersion, and substitution of low-sulfur fuels. During the 1970s, this issue was fought in the courts as well as in Congress, with all three parties—public-interest groups, government, and industry—playing important roles.

One example of the way in which decisions made in court affect a policy formulated in Congress and administered by EPA is the prevention of significant deterioration issue. It began in 1972 when a D.C. district court agreed with the Sierra Club that EPA should refuse to approve any state implementation plan that would allow air quality in any region to become more degraded than it was when the plan was formulated.[66] In the absence of Justice Powell, this decision was upheld by a tie vote of the Supreme Court.[67]

Because the court had used the vague phrase "significant deterioration," however, the decision initiated a new round of policy formulation, first in EPA and later in Congress. After considerable public debate, the EPA in 1974, suggested a compromise solution, dividing all clean air regions into three classes: I, areas of restricted growth, allowing little air deterioration; II, areas of moderate growth; and III, areas where the air quality would be allowed to deteriorate to secondary standards, the level of air quality prescribed for regions classified as polluted. Eventually this tripartite division of the clean air regions of the U.S. was codified by the 1977 amendments to the CAA.[68]

The Sierra Club's suit was not directed solely at electric-generating plants. However, the suit had as great an impact on this industry as on any other stationary source of air pollutants because the strategy of electric utilities has been to move their new plants into the relatively pristine countryside near major metropolitan areas both to escape the high costs of land acquisition in the cities and to avoid emission-control requirements in the polluted areas.

Several other major arguments about the implementation of the CAA have been addressed by the courts, including one over the authority of EPA to control indirect sources of air pollution, such as highways and major shopping complexes. Ultimately many of these arguments were repeated in Congress and resolved by the 1977 amendments, but it is clear that the courts have played a major role in the reformulation of much of the policy with regard to air-pollution control. Nearly all such decisions have important implications for the energy industry. This discussion, however, is restricted to those cases concerning the electric utility industry, disregarding the equally important issue of control of mobile sources of air pollutants.

### *Cases Brought by Environmental Groups*

Before 1973 the few cases involving fossil-fuel-fired power plants were based on laws other than the CAA and made little impact on the industry. For example, in 1972, the Environmental Defense Fund (EDF) succeeded in forcing the Wisconsin Department of Natural Resources to reconsider a permit issued to a coal-fired plant on the Wisconsin River for discharging effluents without sufficient controls to avoid damage to aquatic life.[69] In 1972 a group of New York citizens objected to the water-discharge permit given to the Astoria plant by the Corps of Engineers. The federal district court there found that an EIS had to be written before the Corps of Engineers could allow completion of the plant, but refused to issue an injunction.[70] Later the same court argued that the construction phase of the project was not harmful to the environment, and EPA could control environmental damage through its water-pollution-control-permitting authority after the plant was completed.[71]

In 1972 the Jicarilla Apache Indian tribe, along with several national public-interest groups, including EDF, sued the Department of Interior to force it to write an EIS concerning the Four Corners fossil-fuel-fired plants in New Mexico. However, both the district and U.S. Court of Appeals for the Ninth Circuit found that NEPA requirements had been observed, primarily because the Indian lands had been leased to the energy industry too long for NEPA to apply.[72]

Even after the CAA came into effect and another Indian tribe objected to the same lack of controls on the Four Corners plant, the tribe was told that substituting low-sulfur coal would resolve the problem, and the plant did not need emission controls.[73]

In 1973 an Indian tribe tried to force the Federal Power Commission to regulate power plants that were using the Colorado River for cooling. But the U.S. Court of Appeals for the D.C. Circuit ruled that the FPC has authority only over hydroelectric power and transmission lines, not fossil-fuel-fired plants.[74] The D.C. Circuit, however, was sympathetic to the Indians' plea and asked the FPC to reconsider its authority over projects using water from federal dam projects. But in 1975, the U.S. Supreme Court denied even this authority.[75]

After 1973 most litigation about fossil-fuel-fired plants came to focus on the CAA. A few environmental groups, however, continued to utilize other types of legislation to restrict the industry's development. As late as 1978 a conservation group tried and failed to force EPA to increase the strictness of its standards for a water-discharge permit to a coal-fired plant in the Indiana Dunes.[76] In the same year, the Wyoming Supreme Court refused to reject a permit given a utility by the Wyoming Industrial Siting Council.[77] As in most of the other cases, deference was given to the regulatory agencies' discretion in issuing permits.

By 1973 after the 1970 amendments to the CAA had been in place long enough to be enforced, the number and seriousness of cases brought to court had increased. One of the primary responsibilities of the states under the CAA is to develop state implementation plans (SIPs), which outline the means by which they intend to achieve national primary air-quality standards. The plans were to be submitted to EPA by 1972; the states had three years to achieve their initial clean-up. Provision was also made for a two-year extension, a schedule that would bring dirty areas into compliance with the standards by 1977, and the 1977 amendments extended these deadlines.

When the implementation plans began arriving, the NRDC argued that several of the plans were too weak; its challenges met with varying results. In 1973 the First Circuit accepted several of the NRDC's arguments about the inadequacies of the Rhode Island and Massachusetts plans, ordering EPA either to disapprove the plans or to show how the NRDC objections could be met.[78] The court went so far as to award attorneys' fees to NRDC for performing a public service.[79] In the same year, however, the Tenth Circuit took a much different approach, throwing out a similar suit directed against the SIPs of Utah, New Mexico, and Colorado on the grounds that the NRDC did not have standing since it had suffered no injury.[80] In 1974 the Ninth Circuit effectively agreed with the Tenth when it ejected NRDC from a suit challenging Arizona's SIP. But it went on to the merits of the case because there were Arizona residents involved and found that EPA's approval of the plan was legitimate despite the state's intention to provide for variances from the emission standards created under the SIP, even when such variances would degrade the present clean quality of the air.[81]

The Eighth Circuit also threw out some of NRDC's objections to the Iowa SIP, but it agreed with the First Circuit that the variance procedure was too generous. The court allowed for variances to be issued that would postpone the attainment of the SIP, but it would not allow revisions of the SIP itself.[82] In 1974 the Second Circuit agreed with the First and the Eighth when it reviewed New York's SIP. It remanded the variance plan to EPA for reconsideration, stating that only variances given before the attainment date were legitimate, since these could be considered postponements rather than revisions of the SIP.[83] The Fifth Circuit went furthest in agreeing with NRDC in 1974, when it, like the First, Second, and Eighth circuits, disapproved the variance procedures of the Georgia SIP. It also ruled that the Georgia SIP was inadequate because it allowed for dispersion of emissions through tall stacks rather than by treating pollutants.[84]

In 1975, however, the U.S. Supreme Court overturned the Fifth Circuit's ruling concerning variances. It effectively agreed with the Ninth Circuit that variances could be granted by revising SIPs as long as the primary air-quality standards continued to be met. However, it also ratified EPA's acceptance of the rulings of the First, Second, and Eighth circuits that no post attainment date variances could be granted.[85] Later in 1976 when NRDC tried to obtain a contempt ruling against EPA for allowing Georgia to use the tall-stack dispersion strategy despite the earlier ruling, the Fifth Circuit found in favor of the state and EPA on the grounds that some credit must be given to industry, which had already committed itself to a dispersion strategy before the first court case.[86] The court also refused attorneys' fees to NRDC.[87]

By late 1974 most environmental groups seemed to have withdrawn from the battle over the strictness of SIPs, leaving the field to industry and government. A few minor cases, however, were filed concerning efforts to enforce particular permits. In 1975 a Wisconsin citizens' group sued to force EPA to prosecute Wisconsin Power and Light for violating Wisconsin's SIP. The court found ambiguously that it was up to the state to determine when to prosecute.[88] In 1975 the Sierra Club sued the Nebraska Department of Environmental Control for issuing a permit to build a plant without applying new source-performance standards. The district court allowed the case to be brought despite pleas that it was outside federal jurisdiction.[89] Later, however, when the plant was built, the court allowed the state to permit the plant without emission controls, on the grounds that it had agreed to use low-sulfur coal.[90] This decision, like many of the nuclear cases, rested on the concept of administrative discretion.

In 1976 a federal district court in the District of Columbia gave summary judgment to Friends of the Earth, which objected to a violation of the District's SIP's ban on visible emissions.[91] In 1978 the D.C. Circuit agreed with the Sierra Club that the 1977 amendments did not justify EPA's allowing new sources of air pollution to avoid the best available control technology for emissions reduction because another part of the same plant reduced its emissions.[92]

Even including the all-important cases concerning SIPs, only a modest number of individual fossil-fuel-fired plants were actively challenged by environmental groups during the 1970s. Despite the fact that more of these plants are given permits each year than nuclear plants, the latter type has been much more often opposed. Altogether only twenty-five cases, representing eighteen different controversies, were initiated by public-interest groups either against the government or industry, which included both objections to SIPs and private nuisance suits. (See table 6-1.) Over one-third (nine) of these were decided at least partially in favor of the environmentalists, a larger percentage than in nuclear cases, in which greater deference was shown to the regulatory agency than in nuclear cases. The Supreme Court played a more limited role in these cases. It heard only three on appeal and upheld the most important: the Sierra Club case that created the prevention-of-significant-deterioration concept.[93] In two cases the high court modified the lower courts' earlier rulings, but it did so in a limited fashion.

Two possible explanations may be considered for the differences in treatment of nuclear and fossil-fuel-fired plants. First, the CAA is a precisely worded law in many respects, with much language that favors the environmentalists' pleas. The Atomic Energy Act, on the other hand, was written primarily to promote nuclear power; there is little in it to restrain the AEC or NRC. The vaguely worded NEPA is not very effective as a counterbalance to the broad powers of the Atomic Energy Act. Second, given the complexity of the technology involved in nuclear-power cases, it is understandable that lay judges hesitate to overturn administrative decisions. Judges seem to feel somewhat more confident of their ability to understand the technology involved in the reduction of air pollutants from fossil-fuel-fired stacks and consequently to oversee the expert decisions of the EPA.

### *Governmental Cases*

In part the paucity of cases initiated by persons objecting to individual power plants can be explained by increased activity by other kinds of litigants. In some instances, governmental units have replaced environmentalists by initiating cases against industry and other governmental agencies representing industrial points of view. For example, in 1970 the Los Angeles Water and Power Agency sued the Los Angeles Pollution Control District because it had refused to allow a power plant in the Los Angeles basin. The California trial court decided in favor of the air-pollution agency and prevented the plant from being built.[94] The California Supreme Court decided similarly in a case between the Orange County Pollution Control District and the California Public Utilities Commission, which had given a license to a power plant in another heavily polluted area in 1971.[95]

On occasion state governments have joined environmental groups; for example, the attorney general of New Mexico joined the Sierra Club to object to a Four Corners plant. Originally the trial court agreed to the nuisance suit by the attorney general and the club.[96] The Supreme Court of New Mexico, however, agreed with the utility that the appropriate agency to take action against the plant was the New Mexico Environmental Control Agency. The agency evidently did not wish to prosecute, and there was no remedy available in private nuisance law.[97] Thus it appears that when an administrative agency responsible for enforcing the public law proves less than zealous, older private remedies may be weakened by the existence of the public law.

In addition to conflict between agencies of the same government, there have also been conflicts, as in the nuclear cases, between the federal and state levels. In two similar cases, states sued to force the Tennessee Valley Authority (TVA) to apply for a permit to burn coal in its power plants. In 1973 a federal district court in Kentucky ruled that the TVA did not have to obtain a SIP permit.[98] The next year, the Fifth Circuit ruled that TVA must obtain an Alabama permit for the same type of plant.[99] But the Supreme Court upheld the Kentucky decision.[100]

On occasion state or federal agencies have filed suit against industrial dischargers. In 1971 Wisconsin instituted proceedings against a power company. Unlike the New Mexico Supreme Court, the Wisconsin Supreme Court found that even if the Wisconsin pollution-control agency did not want to prosecute, such reluctance did not preclude the use of a nuisance suit by the attorney general.[101] But when the Chicago pollution-control authorities attempted to use nuisance law to shut down a power plant outside its jurisdiction in Indiana, the Illinois appeals court held for industry, arguing that the city should follow appropriate procedures outlined in the CAA.[102]

In 1973 the appropriate Pennsylvania pollution-control agency used the administrative process to order a utility to control sulfur emissions. The power company argued that since there was no technology available, it could not be held in contempt, and both the trial and appellate courts agreed with it.[103] In 1977 an enforcement suit brought by the U.S. EPA against a utility in Indiana was thrown out by the federal district court there because it considered the emission standards for power plants to be economically and technically unattainable.[104] However, in the same year, a district court in Ohio found in favor of EPA in forcing a city-owned power plant to conform to new plant specifications in the regulations of the CAA.[105]

Altogether there have been only fourteen suits, representing ten controversies, initiated by governmental agencies in order to control emissions from fossil-fuel-fired plants. (See table 6-1.) These include disputes between governmental units and one case in which the state joined with a public-interest group. Since the courts decided eight of these cases in favor of industry, it is obvious that there was no felt need to defer to the expert judgment of the pollution-control agencies involved. The fact that other government agencies were often on the opposite side of the litigation bolstered the courts' willingness to second-guess the pollution-control agency's decision. The one Supreme Court intervention involved federal preemption of state efforts to control air pollution from federal power plants and resulted in less-effective control than the states wanted.

### *Industrial Challenges to Regulations*

The largest number of CAA cases was initiated neither by public-interest groups nor by government but by industry. Some of these cases, however, represent industry response to governmental enforcement actions. In 1971 in two separate cases Pennsylvania courts overturned rulings by the Air Pollution Commission there against coal companies on the grounds that evidence collected by the agency was insufficient.[106] Similarly, in 1971 a utility company successfully appealed its conviction by the New York air-pollution-control agency for allowing fly ash to descend on neighboring communities.[107] In 1971 an Illinois trial court found in favor of the coal industry, which objected to a ban by the Illinois Pollution Control Board on coal use in Chicago.[108] But the appellate court overturned it in 1973, arguing that the control board's order was not final and the industry must exhaust its administrative remedies before filing suit.[109]

It was only after SIPs went into effect in 1972 that a large number of important complaints from industry made their way to the courts. In 1972 the Third Circuit made the first ruling, a procedural one that simply instructed Getty Oil to start its appeal about an SIP in circuit rather than district court.[110] The Fourth Circuit, in 1973, received challenges by utilities to EPA-approved Maryland, Virginia, and West Virginia SIPs. The court ruled that EPA did not have to write an EIS for each SIP. After sending back for a more-complete record from EPA, it agreed that industry had been given sufficient opportunity to comment on the plans at the state hearings.[111] In 1975 the Seventh Circuit also found that EPA did not need to write an EIS before approving SIPs for Illinois and Indiana. Economic and technological questions could be raised later in individual enforcement actions in the district courts when EPA or state agencies began enforcing the SIPs.[112]

By far the largest number of challenges to SIPs was made in the Sixth Circuit where the issue received the most thorough review. In 1973, the Sixth Circuit followed the Fourth's lead in rejecting the idea of an EIS for SIPs, but it also agreed that industry deserved a hearing because of the Administrative Procedures Act. EPA was told to reconsider the Ohio and Kentucky SIPs in light of the arguments of industry.[113] Later, in 1975, the Sixth ducked the issue of whether EPA had sufficiently considered the economic feasibility of sulfur emissions objections, and ruled that the case was not ripe.[114]

In 1975 the Eighth Circuit, in reviewing a challenge to Missouri's SIP, refused to review whether sulfur-oxide emission standards were economically or technically feasible on the grounds that the CAA did not allow such review by courts.[115] In 1976 the U.S. Supreme Court reviewed this case and decided unanimously that EPA could not be forced to reconsider the SIP because of economic contingencies. Chief Justice Burger and Justice Powell wrote a reluctant concurrence in which they argued that Congress should not have passed such strict regulations in the first place because of the economics of energy production. But given the wording of the CAA, there was no way that they could allow the courts to intervene when only EPA and the states had been given authority to consider economic and technical problems.[116]

In 1977 utilities in Ohio perforce shifted their attack to objecting to individual plant emission controls placed on them by Ohio's pollution-control agency. The Cleveland utility sued Ohio's EPA for a variance to its SIP because, it argued, the sulfur-oxides emission standards were unrealistic. The Ohio appeals court remanded the issue to Ohio EPA for reconsideration of the standards but agreed with Ohio's Board of Environmental Appeals that it did not have authority to alter the standards.[117] Dissatisfied with this ruling, the industry appealed to the Ohio Supreme Court to have the entire CAA declared unconsitutional, but the court dismissed for want of a constitutional issue.[118] The U.S. Supreme Court refused certiorari.[119] The conclusion of the state case paved the way for the Sixth Circuit to dispose of a parallel federal case. In 1978 it decided that the Ohio SIP was correctly drafted by EPA, since the Ohio agency was unwilling or unable to come up with a sufficiently strict one. The earlier question raised, but unanswered, in the Buckeye case about the sufficiency of the SIP was decided in favor of EPA, and the Supreme Court again refused certiorari.[120]

The conflict over the Ohio SIP was not over, however, for in 1978 both an environmental group and the industry once more filed suit objecting to the new SIP. Ohio had revised the attainment date from July 1, 1975, to 1977. Although both dates had already passed, each party objected to EPA's discretion to change the deadline. The Sixth Circuit decided that it was within EPA's discretion to make this change and rejected both the environmentalists' and the industry's claims.[121] In 1979 the issue of Ohio's SIP was again raised, and the Sixth Circuit told EPA to reconsider the model it was using to estimate how far emissions are dispersed in rural areas.[122]

In 1975 the Sixth Circuit gave an important victory to EPA over the tall-stacks strategy discussed in 1974 by the Fifth Circuit. The EPA had disapproved

Kentucky's SIP because it allowed utilities to rely on tall stacks instead of emission-control technology, and industry objected. But the court agreed with EPA that "intermittent controls" represented little protection for the environment.[123] It appeared that EPA's own attitude toward tall stacks had been considerably altered by the success of NRDC in challenging such SIPs as Georgia's in which EPA had originally approved a tall-stacks dispersion strategy. Although EPA appealed and the Supreme Court overturned that portion of the Fifth Circuit's ruling criticizing variances, EPA did not bother to appeal the tall-stacks aspect of the Fifth Circuit's decision. The high court refused to grant certiorari to the Sixth's decision about industry's objection to the Kentucky plan, indicating its acquiesence in EPA's new policy.[124]

The Third Circuit repeated nearly all of the steps that the Sixth went through. In 1973 industry sued to force EPA to hold hearings before approving Pennsylvania's SIP, and the court agreed.[125] In 1975 the same company challenged EPA's approval of Pennsylvania's SIP, and the court found for industry on the grounds that EPA had not considered the economic consequences of the requirement that scrubbers be put on coal-fired plants.[126] (This was one of the cases overturned by the dicta of the Supreme Court's Missouri decision.)[127] In 1974 a different Pennsylvania utility protested against the SIP's prohibition against tall-stack disperson strategy. Pennsylvania had given the power company a variance because it did not have the technology to reduce sulfur emissions, but the utility wanted this considered a final solution rather than a postponement of compliance with promulgated standards. The federal district court ruled that such a request belonged in the circuit court.[128] The Third Circuit in 1975 ruled that it had no jurisdiction until EPA saw fit to take enforcement action against the industry, and it would afford industry no relief.[129] The same utility, later in 1976, tried to get the circuit to review EPA's approval of Pennsylvania's SIP, but the circuit found that EPA was not obliged to take economic factors into consideration, following the Supreme Court's Missouri ruling.[130] Finally, in 1978, the EPA began to prosecute this utility for nonconformance to the SIP, and the district court allowed the case to go forward despite the utility's claim that scrubbers were an imperfect technology.[131]

In 1974 after the initial Sierra Club victory concerning the deterioration of clean-air areas, the EPA began enforcing its own definition of prevention of significant deterioration (PSD). This resulted in numerous challenges from industry located in clean areas. The first came in the Sixth Circuit where the court decided that the case should be moved to the District of Columbia, where all issues of this type should be resolved from a national perspective.[132] There the D.C. Circuit upheld the EPA's PSD regulations against a consolidated challenge by utilities from many states.[133] Later in 1977 the D.C. Circuit also refused to review EPA's application of its PSD regulations to a utility in Utah on the grounds that the court need only review standards, not their application.[134]

In 1977 a federal district court in Montana agreed with industry that a plant already planned before EPA began applying new source standards to unpolluted areas should be allowed to exceed PSD regulations because of a grandfather clause in the CAA.[135] However, after the passage of the 1977 amendments to the CAA, which codified most of EPA's regulations concerning PSD, the Supreme Court remanded all cases concerning the applicability of PSD standards to new plants back to their respective circuits for reconsideration.[136] And in 1978, the Fifth Circuit found that EPA was operating within its discretion when it forced a new power plant in Florida to conform to PSD regulations despite the fact that it was planned before they were promulgated.[137]

Altogether some thirty-two cases (representing twenty-one controversies) concerning the application of CAA regulations to the electric utility industry have been initiated by industry against the regulators. The Third and Sixth circuits received a disproportionate number of these cases because of the eagerness of utilities in Pennsylvania and Ohio to litigate their conflicts with EPA and state regulators. They have met with mixed results. Of some nine cases begun in either the federal or state courts in the Sixth Circuit, only two were decided unambiguously in industry's favor. The Third Circuit treated industry better, deciding four out of nine cases in its favor. This judicial victory was offset by the greater zeal with which pollution-control officials in Pennsylvania enforced the law by restricting industrial emissions there. To a degree, the tough attitudes of judges in Ohio must be attributed to the extremely lax attitude of administrative officials in the state and the extreme demands placed on the courts by industry against U.S. EPA there.

Of a total of thirty cases in all eleven circuits, seventeen were decided for the government and thirteen for industry; two cases were decided so ambiguously as to constitute a clear victory for no one. (See table 6-1.) It would appear that when the electric utility industry is the plaintiff in CAA suits, there is a better than 40 percent chance that the courts will treat its demands with some sympathy. This is marginally better than the treatment given environmentalists' complaints against fossil-fuel-fired plants. In marked contrast to the courts' nearly single-minded deference to the NRC, however, it is clear that courts have been willing to overturn EPA decisions regarding fossil-fuel-fired plants, regardless of which side the agency originally favored. Whether this is due to greater deference paid to nuclear regulators as opposed to EPA officials or to the complexity of the CAA is a matter for speculation. Again the Supreme Court played a minor role in the interpretation of the CAA. Although industry appealed many of the decisions made against it, the high court refused to hear most of them.

### *Summary*

Seventy-one fossil-fuel power cases were decided by the courts in the 1970s. (See table 6-1.) Nearly half of these were decided in favor of the environment and the others in favor of power development, regardless of whether the initiator of the case was government, industry, or environmentalists. Primarily because of the eagerness of industry to litigate in Pennsylvania and Ohio, the Third and Sixth circuits have heard a disproportionate number of disputes over fossil-fuel-fired plants. (See table 6-2.) Both the D.C. and Seventh circuits also processed a number of these cases, but these caseloads are more in line with the normal large number of cases processed by these two circuits. Unlike the D.C. Circuit in the nuclear-power cases, no one circuit heard nearly half of all the fossil-fuel cases.

The more favorable attitude of courts toward the environment in fossil-fuel cases is due partly to the role that EPA has played. It has frequently represented the environmental interest, as opposed to the NRC, which has not varied from its role as surrogate for industry in nuclear plant disputes. The EPA has shown a willingness to learn from public-interest attorneys and to use their arguments later against industry. After major court victories for public-interest groups, EPA changed from a relatively pro-utility stance to a more pro-environment posture, both with regard to the prevention of significant deterioration of clean areas and with regard to tall-stack dispersion strategy. The tripartite division of labor among industry, government, and environmentalists has aired many issues. But the tough wording of the CAA has enabled (or forced) courts to agree to strict adherence to environmental guidelines. The Supreme Court, in contrast to its behavior with regard to nuclear regulation, has kept out of the interpretation of the CAA to a large degree. In those five cases where it has acted, it upheld the CAA's provisions in three cases.

This analysis does not mean that fossil-fuel-fired plants are more closely regulated than are nuclear plants. In fact, it is often argued that just the reverse is true and that nuclear regulations are tighter because of the newness and riskiness of the technology. Given the relatively small percentage of fossil-fuel-fired plants that have been challenged in court, the obsolescence of many of these plants, which have remained largely unaffected by the CAA thus far, and the refusal of a major portion of the industry to adopt scrubber technology even for new plants, it is clear that the electric-utility industry will remain a major contributor to the air-pollution problem in this country for many years. When cases do get to court, however, the chances for a verdict favoring the environment over power production is greater in fossil-fuel cases than in nuclear ones. The courts' willingness to defer to administrative technical expertise is much less in evidence than in nuclear cases. In those rare cases where federalism became an issue, the court tended to favor the central government, just as in nuclear cases.

## Conclusions

One problem that courts are not designed to handle is the trade-offs between one type of pollutant and another. Courts frequently balance the equities between economic and environmental costs in individual cases. But each decision is rendered on an ad hoc basis, and there is no overall consideration of the cumulative balance of all costs involved. There can be even less attention given to the relative costs of different forms of energy and their cumulative impact on the environment by the courts. Such calculations, if they are to be made anywhere, must be decided in the Congress or in the executive branch. To date there is little evidence that anyone has been able to bring together objective comparisons of such costs and benefits.

Very little is known about the objective costs and the relative impact of the different forms of energy. Most of the available data come from advocates or opponents of one form of energy or another. In the case of nuclear power, we are only beginning to comprehend the enormity of the costs still to come from the disposal of radioactive wastes now housed in holding tanks. And no effort has been made to include in calculations by nuclear utilities the federal taxpayers' contribution to the research and development effort necessary to make the "peaceful atom" possible. Yet claims for the relative cheapness of this form of energy were the primary reason for committing ourselves to the nuclear option.

On the other side of the argument, we may know equally little about the true costs of the fossil fuels we use. If there is one thing that became evident to drivers in the late 1970s, it was that neither government nor consumers had any notion of what they would be charged for petroleum in the future. The total costs of coal production and conversion to electricity also remain a mystery. At present the fossil-fuel industry remains locked in battle with EPA over the costs of environmental clean-up. One of the major weapons of each is the "facts" about the real costs of restoring strip-mined land and technology necessary to clean sulfur and fine particulates from the air. As long as each side uses its information as a weapon in an adversarial situation, there can be little objective comparison of the real costs of nuclear power as opposed to fossil fuels.

The unknowns dealt with here are the easy part of the equation; they concern the costs of cleaning up, the costs of regulation, the costs of power conversion. They have nothing to do with the more difficult part of the equation with which we as citizens ought to be the most concerned; the costs to the public health and safety from each type of power production. These kinds of questions remain even less amenable to objective answers.

Judges, like other lay persons, are caught in the dilemma of having to make specific decisions about particular power plants without nearly the information they need to make such decisions. Yet it would be unwise to argue that such individuals should be excluded from the policy-making process. Given the inability of most experts to agree about the real costs of any given type of energy-conversion process and the willingness of experts to use their own figures in competition with opponents, it seems only reasonable to include as many nonexpert adjudicators in this controversy as possible. Courts may not always have made wise decisions about individual power plants. It is doubtful that any given set of judges would agree about the optimal use of different types of energy any more than administrators or industry spokespersons would. It also seems useful to maintain the judiciary as a court of last resort for such decisions.

## Notes

1. 42 U.S.C. 4321-4347.
2. 42 U.S.C. 7409 et seq.
3. 30 U.S.C. 1201 et seq.
4. New Hampshire v. AEC, 1 ERC 1053.
5. Northern States Power v. AEC, 320 F Supp 172.
6. Northern States Power v. AEC, 447 F 2d 1143, 3 ERC 1976.
7. Commonwealth Edison v. Pollution Control Board of Illinois, 4 ERC 1303.
8. United Mine Workers Union v. Colorado PUC, 1 ERC 1115.
9. Consolidated Edison v. Hoffman, 11 ERC 1346.
10. Seadade v. Florida Power and Light, 1 ERC 1146, 2 ERC 1223.
11. United States v. Florida Power and Light, 311 F Supp 1391.
12. Thermal Ecology v. AEC, 2 ERC 1405, 2 ERC 1379 (433 F 2d 524).
13. Lloyd Harbor v. Seaborg, 2 ERC 1381.
14. Calvert Cliffs Coordinating Committee v. AEC, 449 F 2d 1109.
15. Izaak Walton v. Schlesinger, 337 F Supp 287.
16. Minnesota Citizens Association v. AEC, 4 ERC 1877.
17. Coalition for Safe Nuclear Power v. AEC, 463 F 2d 954.
18. Brooks v. AEC, 476 F 2d 924.
19. Union of Concerned Scientists v. AEC, 499 F 2d 1069.
20. York Committee for Safe Environment v. NRC, 527 F 2d 812.
21. Citizens for Safe Power v. NRC, 524 F 2d 1291.
22. North Anna Environmental Coalition v. NRC, 435 F 2d 655.
23. Concerned Citizens of Rhode Island v. NRC, 430 F Supp 627.
24. Mid-America Coalition for Energy Alternatives v. NRC, 12 ERC 1718.
25. Ecology Action v. AEC, 492 F 2d 998.
26. Natural Resources Defense Council v. NRC, 547 F 2d 633.
27. Aeschliman v. NRC, 547 F 2d 622.
28. Vermont Yankee Nuclear Power Corporation v. NRDC, 97 SC 1098.
29. Sierra Club v. Hickel, 467 F 2d 1048.
30. Gage v. AEC, 479 F 2d 1214.
31. Gage v. Commonwealth Edison, 356 F Supp 80.
32. Seacoast Anti-Pollution League v. Costle, 572 F 2d 872.
33. Public Service Company of New Hampshire v. NRC, 582 F 2d 77.
34. Carolina Environmental Study Group v. AEC, 510 F 2d 796.
35. Carolina Environmental Study Group v. AEC, 431 F Supp 203.

36. Duke Power Company v. Carolina Environmental Study Group, 98 SC 2620.

37. Colorado PRIG v. Train, 373 F Supp 991.

38. Colorado PRIG v. Train, 507 F 2d 473.

39. Train v. Colorado PRIG, 426 U.S.1.

40. Porter County Chapter of the Izaak Walton League v. AEC, 516 F 2d 513.

41. Northern Indiana Public Service Company v. Porter County Chapter of the Izaak Walton League, 423 U.S. 12.

42. Porter County Chapter of the Izaak Walton League v. AEC, 533 F 2d 1011.

43. Committee for Nuclear Responsibility v. Seaborg, 461 F 2d 783, 463 F 2d 788, 463 F 2d 796.

44. Patsy Mink v. EPA, 464 F 2d 742, and EPA v. Mink, 410 U.S. 73.

45. Morningside Renewal Council v. AEC, 482 F 2d 234.

46. Morningside Renewal Council v. AEC, 94 SC 3080, 417 U.S. 951.

47. United States v. City of New York, 463 F Supp 463.

48. Crowther v. Seaborg, 312 F Supp 1205.

49. Sierra Club v. AEC, 6 ERC 1980.

50. Scientists Institute for Public Information v. AEC, 481 F 2d 1079.

51. East Tennessee Energy Group v. Seamans, 7 ERC 2144.

52. Nader v. Ray, 363 F Supp 946.

53. Nader v. Ray, 513 F 2d 1045.

54. Honicker v. Hendrie, 12 ERC 1700, and Honicker v. NRC, 590 F 2d 1207.

55. Natural Resources Defense Council v. NRC, 539 F 2d 824, 9 ERC 1414.

56. NRDC v. NRC, 97 SC 1578, 430 U.S. 944.

57. New York v. NRC, 550 F 2d 745.

58. Garrett v. NRC, 11 ERC 1567, 11 ERC 1685.

59. NRDC v. Energy Research and Development Administration, 451 F Supp 1245.

60. Vermont Yankee v. NRDC.

61. NRDC v. NRC, 582 F 2d 166.

62. Illinois v. NRC, 591 F 2d 12.

63. Lettie McSpadden Wenner, *One Environment under Law* (Pacific Palisades, Calif.: Goodyear, 1976), p. 8.

64. Nielson v. Seaborg, 348 F Supp 1369.

65. Carolina Environmental Study Group v. AEC; Duke Power Company v. Carolina Environmental Study Group.

66. Sierra Club v. Ruckelshaus, 344 F Supp 253.

67. Fri v. Sierra Club, 412 U.S. 541.

68. Clean Air Act, 41 USC 7401, sec. 127.

69. Environmental Defense Fund v. Department of Natural Resources, 4 ERC 1045.

70. Citizens for Clean Air v. Corps of Engineers, 349 F Supp 696.
71. Citizens for Clean Air v. Corps of Engineers, 356 F Supp 14.
72. Jicarilla Apache Tribe v. Morton, 471 F 2d 1275.
73. Oljato Chapter v. Train, 515 F 2d 654.
74. Chemehuevi Tribe v. Federal Power Commission, 489 F 2d 1207.
75. Chemehuevi Tribe v. Federal Power Commission, 420 U.S. 395.
76. Porter County Chapter of the Izaak Walton League v. Costle, 571 F 2d 359.
77. Laramie River Council v. Industrial Siting Council, 12 ERC 1769.
78. NRDC v. EPA, 478 F 2d 875.
79. NRDC v. EPA, 484 F 2d 1331.
80. NRDC v. EPA, 481 F 2d 116.
81. NRDC v. EPA, 507 F 2d 905.
82. NRDC v. EPA, 483 F 2d 690.
83. NRDC v. EPA, 494 F 2d 519.
84. NRDC v. EPA, 489 F 2d 390.
85. Train v. NRDC, 421 U.S. 60.
86. NRDC v. EPA, 529 F 2d 755.
87. NRDC v. EPA, 539 F 2d 1068.
88. Wisconsin's Environmental Decade v. Wisconsin Power and Light and EPA, 395 F Supp 313.
89. Sierra Club v. Train, 7 ERC 2030.
90. Sierra Club v. Train, 11 ERC 1173.
91. Friends of the Earth v. PEPCO, 419 F Supp 528.
92. Sierra Club v. EPA, 11 ERC 1129.
93. Sierra Club v. Ruckelshaus; Fir v. Sierra Club.
94. Los Angeles Water and Power Board v. Los Angeles Air Pollution Control District, 1 ERC 1580.
95. Orange County v. Public Utility Commission, 2 ERC 1602.
96. State of New Mexico, Sierra Club et al. v. Arizona Public Service Company, 3 ERC 1617.
97. State of New Mexico, Sierra Club et al. v. Arizona Public Service Company, 5 ERC 1385.
98. Commonwealth of Kentucky v. Fri, 362 F Supp 360, 497 F 2d 1172.
99. Alabama v. Seeber, 502 F 2d 1238.
100. Hancock v. Train, 96 SC 2006, 426 U.S. 167.
101. Wisconsin v. Dairyland Power, 2 ERC 1763.
102. Chicago v. Commonwealth Edison, 7 ERC 1480.
103. Pennsylvania v. Pennsylvania Power, 5 ERC 1373, 6 ERC 1328.
104. United States v. Public Service Company of Indiana, 12 ERC 1495.
105. United States v. City of Painesville, 431 F Supp 496.
106. Bortz Coal Company v. Air Pollution Commission, 279 A 2d 388; North American Coal v. Air Pollution Commission of Pennsylvania, 2 ERC 1754.

107. Niagara Mohawk Power v. Mosher, 3 ERC 1481.
108. Roth Adam v. Pollution Control Board of Illinois, 4 ERC 1189.
109. Roth Adam v. Pollution Control Board of Illinois, 5 ERC 1145.
110. Getty Oil v. Ruckelshaus, 342 F Supp 1006, 467 F 2d 349.
111. Appalachian Power Company v. EPA, 477 F 2d 495, 579 F 2d 846.
112. Indiana and Michigan Electric Company v. EPA, 509 F 2d 839.
113. Buckeye Power v. EPA I, 481 F 2d 162.
114. Buckeye Power v. EPA II, 525 F 2d 80.
115. Union Electric v. EPA, 515 F 2d 206.
116. Union Electric v. EPA, 427 U.S. 246.
117. Cleveland Electric Illuminating Company v. Williams, 12 ERC 1081.
118. Cleveland Electric Illuminating Company v. Williams, 12 ERC 1087.
119. Cleveland Electric Illuminating Company v. Williams, 12 ERC 1088.
120. Cleveland Electric Illuminating v. EPA, 572 F 2d 1150, 12 ERC 1156.
121. Northern Ohio Lung Association v. EPA, 11 ERC 1411.
122. Cincinnati Gas and Electric Company v. EPA, 578 F 2d 660.
123. Big Rivers Electric Company v. EPA, 523 F 2d 16.
124. Big Rivers Electric Company v. EPA, 8 ERC 2164.
125. Duquesne Light v. EPA, 481 F 2d 1.
126. Duquesne Light v. EPA, 522 F 2d 1186.
127. Union Electric v. EPA, 427 U.S. 246.
128. West Penn Power v. Train, 378 F Supp 941.
129. West Penn Power v. Train, 522 F 2d 302.
130. West Penn Power v. Train, 538 F 2d 1020.
131. United States v. West Penn Power, 460 F Supp 1305.
132. Dayton Power and Light v. EPA, 520 F 2d 703.
133. Public Service Company of Colorado et al. v. EPA, 540 F 2d 114.
134. Utah Power and Light v. EPA, 553 F 2d 215.
135. Montana Power Company v. EPA, 429 F Supp 683.
136. Montana Power Company v. EPA, 97 SC 1597.
137. Gulf Power Company v. EPA, 11 ERC 1805.

# 7 Power and the Environment: A Case Study of the Siting of a 765 Kilovolt Transmission Line

*Karen Burstein*

Theoretically there is a straightforward, comprehensive answer to the present energy-environment dilemma: we must shift from our current dependence on fossil fuels to reliance on renewable and clean resources. Uranium produced by breeder reactors meets the first criterion of renewability, but it is dangerous and costly. The only technologically feasible and economically rational alternative, therefore, is a dispersed, multimodal solar system using biomass, wind, wood, the sun, and water.[1]

In 1979 Barry Commoner delineated quite specifically the choices we should make and the steps we should take to effect the transition.[2] He pointed out the practicality of and return on investment by the Department of Defense in photovoltaic cells, the methane potential in America's agricultural land, and the pumped storage opportunities presented where good wind regimes exist. He argued persuasively that natural gas can serve to bridge the distance between today and tomorrow.

What we lack, according to Commoner, is the national will to confront the fact that no invisible hand will build that bridge; that private oil and utility companies have a substantial stake in aborting the solar future; that concentration of capital, centralization of operation, and economies of scale are mostly inapposite or irrelevant to efforts to harness the sun's power, and that "social governance," governmental intervention on behalf of the real interests of the people, must supplant our lingering belief in the principles of Adam Smith.[3]

Commoner's vision is compelling and his logic consistent. I appreciate his presentation of a blueprint for action. However, as a public utility regulator, I find that blueprint insufficiently particularized to fit the policy constructs with which I must work. Part of the difficulty is, admittedly, the limitations on my agency's authority and the diffusion of energy and environmental responsibilities among a variety of sometimes-competing units—within the state and federal governments. But part is attributable to the fact that the questions that Commoner suggests confront us refer to too global a universe of conflict. At that level of generality, choices are relatively too simple to make and costs, benefits, and risks relatively easy to identify and weigh. In the real world where daily decisions accumulate, the problems that need resolution are considerably more knotty and grubby, the values less clear-cut, available analysis less precise.

My proposition is graphically illustrated by a case that the New York State Public Service Commission concluded in 1978. It began in 1973 when the Power Authority of the State of New York (PASNY) filed an application to construct a 765 kilovolt (kV) transmission line to bring Canadian-produced hydroelectric power to New York City. The case raised issues generally ignored in most debates on energy and the environment.

## Structural Background

Electric consumers in New York State are served by seven private utility companies and PASNY.[4] The private companies are regulated by the Public Service Commission (PSC), an independent agency whose seven members are appointed by the governor and confirmed by the Senate to terms of six years. The PSC basically is responsible to ensure that customers receive safe and adequate service at just and reasonable rates.[5]

In 1970 the legislature enacted Article VII to the Public-Service Law, which requires the PSC to issue a certificate of environmental compatibility and public need before any major utility transmission line is sited. The statute defines a major electric transmission facility as one with a design capacity of at least 125kV extending for one mile or more.[6] The commission's jurisdiction includes facilities that PASNY proposes to build.

PASNY closely resembles the federal Tennessee Valley Authority. It is run by a board of trustees, nominated by the governor and confirmed by the Senate to terms of one to five years. The governor selects the chair and the board appoints all necessary employees.[7]

In the 1950s PASNY developed the hydroelectric potential of Niagara Falls and the St. Lawrence River, and in 1968, it engaged in planning nuclear installations designed to improve the state's base-load generating capacity.

Until 1972 only upstate communities benefited from its operations, since statutes limited its power distribution to "economic transmission distances."[8] Its primary customers were municipal and rural electric cooperatives. However, because the Niagara project replaced a collapsed dam originally constructed by a private utility company, PASNY also served some of that utility's industrial and residential users.

In 1972 following its declaration of a power-capacity shortage in southeast New York, the legislature instructed PASNY to supply electricity to the transportation authorities of New York City. In 1974, directed to purchase two uncompleted plants of the financially troubled Consolidated Edison Company (Con Ed), PASNY was further authorized to sell energy directly to New York

City and to municipal corporations, like Westchester County, within the city's metropolitan region.

PASNY is financed by the issuance of tax-free bonds and notes. Within the broad ambit of enabling legislation and the bond covenants it writes, it is largely self-regulating.[9] It sets its own rates, service classifications, and practices. It must obtain prior approval from the PSC before constructing and operating a major transmission line, but by virtue of both Article VII of the Public Service Law and the authority's governing statute, PASNY makes the threshold determination that need for such facility exists.[10] That determination is conclusive on the PSC.

## The Case

In 1973 PASNY entered into a contract with Hydro-Quebec to purchase 800 megawatts of firm power during the months of April to October in the years 1978 through 1996.[11] The contract contemplated that additional unspecified amounts of nonfirm diversity power might later be available. In September 1973, PASNY applied for a certificate to build and operate a 765kV line of some 155 miles in length, extending south from the border between New York and Canada to the town of Marcy near Utica.[12] It asserted that the line was necessary to deliver energy that it would be importing to downstate customers.

Apart from the fact that its declaration of need bound the commission at the outset, this application differed in significant respects from ones that the PSC has previously considered and indeed those it would consider in the future. Ordinarily transmission lines are conceived as extensions of generation planned by utilities or as means of increasing the reliability of their electric supply to the people in their franchise territory. Here the applicant was suggesting a facility to connect with an installation that had been developed by another country. Moreover, it was offering power to an area outside of its previous service responsibility because of a 1972 legislative mandate.[13] It intended to move that power, produced by a hydroelectric plant, not for reliability purposes but to replace downstate oil-fired generation. Finally, the capacity of the line was the largest ever proposed for the state; until the time of this application, nothing greater than 345kV had been built.

Given these factors, it is not surprising that a significant variation from the ordinary process of handling applications would occur. In the normal course of events, that process is relatively straightforward. The PSC appoints an independent administrative law judge to preside over an evidentiary hearing in which aesthetic, ecosystem, and land-use-impact issues are tried.[14] The applicant's case is cross-examined by the staff of the PSC, the departments of Health, Agriculture, and Environmental Conservation, and other affected parties that have a statutory right of participation.[15] They in turn may offer alternative direct cases for cross-examination by the applicant and each other. Briefs summarizing the various positions are submitted to the judge, who writes a recommended decision, which generates further clarifying briefs. The entire record then goes to the PSC for final action. The kinds of questions with which such proceedings normally grapple concern the preferred and alternate routes of the applicant; the relative economic impact on particular parcels of land affected by the location of a right-of-way; the necessity of avoiding ecologically sensitive areas like wetlands or wild forests; the nature and aesthetic intrusions of structures that support the lines; and the vegetation management plans that the applicant intends to implement.

The PSC's determination to grant a certificate must follow its findings on the basis of this material, that need exists for the facility (except, of course, in the case of PASNY); the facility represents the minimum adverse environmental impact, considering the state of available technology and the nature and economics of the various alternatives; some, if any part, should be undergrounded; the line conforms to the long-range plans for expansion of the electric power grid of the electric systems serving the state; the location of the facility conforms to applicable state and local law, unless unduly restrictive; and the facility will serve the public interest, convenience, and necessity.

Unlike a rate case, which must conclude within eleven months, an Article VII proceeding has no time limitations.[16] Because of the comprehensiveness of the findings, which ultimately underpin a PSC ruling, and because the line proposed is almost always projected to be built several years in the future, the hearings occur at a pace more leisurely than frenetic. Thus in the summer of 1974, almost a year after the filing of the application by PASNY, the first evidence was considered.

The initial divergence from customary practice occurred in November 1974 when the departments of Health, Agriculture, and Environmental Conservation informed the PSC that articles in medical journals and reports of Soviet experience with 765kV facilities suggested certain health and safety problems associated with extra-high voltage lines: notably audible noise, induced current shocks, and biological effects of magnetic and electric fields.[17] The PSC, through the examiners, agreed that these issues demanded extensive exploration and, again diverging from regular practice, severed them from questions of routing and land-use impact. However, in so doing, the PSC expressed concern that the certification process not be unduly extended. It thus required that the original case go forward while the new one on health and safety got underway and that the record in the latter be directly certified to it as evidence accumulated.[18]

## What the Commission Did

The hearing on the route for the line took forty-one days and the testimony of twenty-eight sworn witnesses. On June 28, 1975, the administrative law judge recommended certification of the facilities along a modified route. By October 1975, the record, briefs, and recommended decision on that question had reached the PSC. Since the safety and health proceeding was still pending and had not been taken into account in the material before the PSC, the commission instructed the judge to solicit further testimony on whether partial certification would prejudge a final determination on health and safety issues. By February 6, 1976, the PSC felt able to authorize the beginning of right-of-way clearance and access-road construction at nine specific locations covering 122 miles.[19]

The commission rested its decision on three grounds, explicitly identifying benefits that the line would confer. First, by this time PASNY had offered uncontroverted evidence that the electricity produced by Hydro-Quebec could be delivered to New York City at about half the cost of that generated by existing oil-fired plants. PASNY estimated that for each year construction was delayed, downstate consumers would pay an additional $45 million in fuel charges and would incur $16 million in costs for interest during construction and projected inflation. Second, to the extent that imported power replaced domestic fossil generation, there would be environmentally beneficial reductions in particulate, nitrogen, and sulfur-oxide emissions. Finally, PASNY claimed that the partial certification of the preliminary route-preparation work could save from four to six weeks of construction time on the line.[20]

These benefits were weighed against the risks that partial certification posed. The PSC concluded that such risks were slim because it was convinced that the proposed steps under consideration would be necessary no matter what voltage was ultimately authorized, because the area to be cleared was sparsely enough populated not to preclude later widening of the right-of-way, and because the route and land configuration was equally desirable for emplacement of a facility of 345kV design capacity.[21]

This order of February 6, 1976, actually conditioned partial certification on PASNY's notifying the commission within twenty days that it would be willing to build and operate a 345kV line if higher voltages were found not to be in the public interest.[22] PASNY replied with a statement that it "certainly intend[ed] to build a line to Quebec."[23] By subsequent order, the commission indicated that it was reading this equivocal language to mean compliance, unless directly and immediately controverted.[24] PASNY maintained silence.

On the basis of its benefit-risk analysis, the PSC said that the application for additional transmission facilities along the general route proposed by PASNY conformed to the long-range plan for expansion of the state's electric power grid; the nature of the probable environmental impact of the limited clearing had been properly evaluated; the corridor and route segments authorized represented the minimum adverse environmental impact; they conformed to applicable local and state law; and the route would serve the public interest and convenience. In the order, the commission further declared that undergrounding of a 765kV line was not technically feasible and was economically unwarranted in the case of 345kV.[25]

On June 30, 1976, the PSC authorized construction on the previously certified route, although the health and safety hearings had not yet concluded.[26] Two significant events had occurred since its partial certification order of February 1976. First, all direct testimony on hazards associated with extra-high voltage lines had been submitted, and almost all had been cross-examined. Second, without approval and in contravention of the commission's February understanding, PASNY had purchased structural equipment—in particular conductors—that could be used only with a line of 765kV configuration.

From a procedural point of view, neither of these events really required response from the commission. However, various pressures militated against inaction. It is quite clear from the opinion of June 30, 1976, that the commission was bent upon finding an intellectual justification for making a decision. How it managed deserves some detailed discussion.

The PSC began by saying that it could not authorize construction unless it found (1) that operation of the lines at the contemplated voltages with the support structures, conductors, and clearances authorized presented no significant health or safety problems, or (2) that the opportunity to eliminate any health or safety threats would not be foreclosed by permitting construction to proceed at this time.[27] The commission implicitly acknowledged that it could not find question 1 to be true. However, question 2 offered considerably more leeway.

The commission started its treatment of question 2 by analogizing the situation presented to a motion for summary judgment.[28] Movants in such cases say that assuming the truth of all the factual allegations raised by the other side, relief on my behalf is proper. Here the commission recognized that it had on record the statement of the worst possible hazards.[29] Accepting that they all existed, it asked whether they were sufficient to preclude its authorizing construction right away.[30] On the basis of what it denominated a worst-case analysis, it concluded negatively. The PSC then summarized the hazards at issue and the suggested responses to them.

Among the most significant hazards were the biological effects of electric and magnetic fields. The passage of an electric current through any unshielded conductor of all overhead transmission lines produces both electric and magnetic fields in the surrounding medium. Witnesses testified that tower design and conductor configuration or size had little or no impact on the intensity of the field produced, that such intensity is attenuated with increasing rapidity the farther one moves from the facility, and that some laboratory studies indicate that extended exposure to such fields might have neural or behavioral effects and possibly other biological consequences.[31] No witness claimed that this evidence required a construction prohibition; rather suggestions for mitigation of possible problems ranged from undergrounding to widening the right-of-way to ordering research on the long-term, subtle consequences.[32]

A second hazard was audible noise. 765kV lines differ from those of lower voltage because they produce audible noise during periods of inclement weather. However, at the edge of the right-of-way, under the worst conditions, this noise is approximately as loud as normal conversation. Although it might interfere with sleep or speech, there was clear evidence that it would not cause temporary or permanent damage to ears. Testimony showed that noise effects could be mitigated in two ways. The first required conductors of a different design from those PASNY had acquired. The other meant establishment of a protective zone beyond the ordinary right-of-way wherein authorities could forbid the construction of new residences and order the purchase or removal of existing homes.[33]

The third most significant hazard was induced shock. A 765kV line, like all other overhead transmission facilities, induces an electric charge in insulated conducting objects near the line. Anyone who is standing on the ground and touching such an object will receive two kinds of shock: a steady-state or short-circuit current and a transient current or spark discharge. Direct physical harm from a short-circuit current occurs only above the "let-go" level, the point at which involuntary muscle contractions prevent an individual from releasing the conducting objects. The PSC staff testified that the safe "let-go" level for men is about 9 milliamperes (ma), for women about 6ma, and for children about 4.5ma. The Russians, with considerable experience with extra-high voltage lines, say the current should be held to 4.0ma.

Currents are less, the greater the transmission line conductor to ground clearance, the less the degree of insulation of the conductive object, and the effective grounding of persons who touch it.

Since PASNY was proposing ground-to-conductor clearances ranging from sixty-three feet over public roads to forty-eight feet over areas other than public or private roads, tests were conducted outdoors with various artificially insulated vehicles and artificially grounded subjects to achieve a worst-case simulation. At the forty-eight-foot clearance, it was impossible to exceed the safe "let-go" level for men and only a very large vehicle, like a tractor-trailer, provided a current substantially in excess of the Russion-literature-suggested 4.0ma.

The commission decided that the worst case—which posited a very large tractor-trailer, parked precisely in the right place, under a maximally loaded line at the fullest possible voltage, being touched by a well-insulated child—was much too remote to be determinative. Instead the PSC relied on a more-probable worst case, premised on assumptions of what vehicles would ordinarily be found under the line at different clearances. It concluded that in this simulation, expected current levels did not exceed 1.0ma.

The commission justified its approach by noting that field experience with high-voltage lines in Canada and the territory of the midwestern American Electric Power Co. (AEP) had yielded no reported incidents of direct or indirect harm due to shock. Further, the clearances of the PASNY line were more conservative than those used by the Canadians, AEP, and the Soviet Union.[34] Finally, the minimum clearance would occur only under extraordinary circumstances, requiring full loading of the line (4,000 megawatts) at the maximum voltage (800kV), particularly unlikely since transmission line voltages tend to decrease as loadings increase.

Although the commission's analysis of electric current shock is intellectually supportable, it does occur, as one dissenting member noted, in the context of a realization that PASNY's material purchases had foreclosed the option to raise conductor heights.[35] The commission glossed over this fact, saying that it could resolve most induced current difficulties by bonding and grounding efforts, a statement entirely true only about stationary objects.[36]

In summary, the commission's review of the health and safety hearing data, by application of worst-case analysis, consistently ended with the expressed conviction that the PSC retained adequate flexibility to impose necessary operating conditions after the line was constructed. So with respect to electric and magnetic fields, it could widen the right-of-way and undertake further experimentation on long-range effects; in the matter of audible noise, it could establish an additional protective zone; and in the matter of induced shock, it could ensure that lines were strung no lower than PASNY proposed and create a bonding and grounding program.

Once more, the PSC was weighing costs against gains. Since the risks it had identified seemed to it capable of amelioration before the line was energized, they seemed less significant than did the benefits the line promised. Again adducing fuel savings, pollution diminution, and the avoidance of interest costs, it pointed out that without construction authorization, the line could not be completed in time to take power under the contract in 1978. It noted that the 2.7 million customers buying electricity from Con Ed represented more than half of the state's population and paid the highest rates in the country. The PSC contrasted these with the rates of upstate consumers who were able to take advantage of the cheap energy produced by PASNY's St. Lawrence and Niagara installations. The commission underlined the fact that many people in Con Ed's area had incomes below the poverty line, were unemployed, and had to make painful choices among lights, food, and medicine.[37]

Finally, in an almost embarrassed aside, it acknowledged that the threshold determination of need was beyond its jurisdiction in any event and admitted that the savings to be realized by proceeding, relevant to the comparative evaluation of all public interest factors under Article VII, "are available only because the Power Authority has apparently gone ahead without authorization from the PSC and purchased structural equipment required for the construction. Any substantial redesign of the line would require at least an additional six months and the lead time on the purchase of structural steel has been estimated to be as long as 18 months."[38]

Moreover, the PSC asserted that should further evidence emerge indicating that the structures did have to be redesigned, the commission reserved its rights to order them torn down. However, on the basis of what it then knew, it re-affirmed that "there is clear evidence of the positive economic advantages of the 765kV line" and "there is nothing in the worst case already before us to justify the conclusion that such line should not be constructed."[39]

## The Case Concludes

Notwithstanding PSC protestations that it retained considerable flexibility, the June 30, 1976, decision effectively determined the final outcome of the case. For the next two years, various other portions of the line were cleared and constructed.[40] On June 19, 1978, the commission issued its concluding order.[41] The hearings on health and safety produced more than fourteen thousand pages of testimony from thirty-one expert witnesses and close to 150 exhibits. It is probably fair to say that nothing in that record substantially contradicted the information available in June 1976, but its weight persuaded the PSC to take some steps only contemplated then.

PASNY had orginally proposed a right-of-way of 250 feet. The applicant said that testimony was too weak to conclude that exposure to the field produced posed a sufficiently unreasonable risk to human health to require further regulation by the commission. The PSC demurred, arguing that "the weight of scientific evidence in the case dictates that observed effects not known to be benign must be considered potentially hazardous unless they are temporary or reversible."[42] Because the applicant had not refuted the inference of harm, the PSC determined that the right-of-way be extended to 350 feet so that the field intensity at the outer edge would be equivalent to that at the edge of normal right-of-ways for 345kV lines–"with which we have lived for two decades."[43] Acknowledging that experience with 345kV proved only the absence of gross impacts and focusing on the subtle effects of chronic exposure, it further ordered that a properly designed program of study be undertaken over the next several years to explore questions raised in the case. The study was to be funded by PASNY and other utilities and directed by independent researchers under PSC supervision.[44] Pending the study's outcome, no higher fields would be permitted and within the right-of-way, no houses or recreational uses would be allowed.

The commission, granting that PASNY had volunteered to begin a grounding program, mandated one anyway and added an educational effort to acquaint people likely to enter the right-of-way with precautions that they could take to reduce shock annoyance.

Finally, after discussing at length the various sound frequencies that should be guarded against, the commission settled on 35 decibels as the proper maximum noise level for bedrooms. Because response to noise is so subjective, the PSC directed PASNY to report to the commission and attempt to resolve all complaints concerning audible noise. If the complaint were made by a home owner within a zone extending six hundred feet from the center line of the certified route and it could not be informally resolved, the PSC might, within eighteen months of the line's operation, require PASNY to purchase or remove the house.

PASNY's response to the conditions that the commission imposed was to take the PSC to court. At present, there is a three-to-two decision from the appellate division affirming the right of the PSC to establish a 350-foot right-of-way and a zone of protection against noise invasion.[45] However, the court found that the PSC attempt to resolve the noise problem by compelling home purchases was outside the authority granted it by statute, as was its attempt to assess PASNY for a study on long-term health effects of high-voltage lines. According to the majority, however, the commission could have required such a study before certification under Article VII. The minority dissented, asserting that the study and noise remedies were clearly within the contemplation of the legislature when it vested exclusive jurisdiction over line siting in the commission. While the commission might have appealed this decision as of right, both the PSC and PASNY delayed further judicial review with the view toward negotiating an acceptable settlement.[46]

### What the Commission Could Have Done

The commission majority contends that had there been no preliminary and partial certification orders, the end product would have been substantially the same, except that downstate consumers would have borne some additional costs. However, as the dissent in the June 30, 1976, and sequential orders point out, the due-process violation—making worst-case analysis before the actual hearing concluded—may have dangerous precedential consequences for the future and, to some degree, impairs the integrity of the PSC's own posture. Under the theory of the dissenting Commissioner, the PSC should have ignored allegations of savings (which he pointed out amounted to $4.92 annually per average customer) and PASNY's unauthorized purchasing activity and allowed the health and safety case to finish before reaching a determination.[47]

Regardless of whether the majority is correct about the final outcome, the dissent represents the better approach. Procedural regularity would not only have avoided potential institutional damage; it would have permitted closer examination of the full record's implications. Instead the PSC's premature decisions prevented its consideration of a number of alternative solutions.

For example, it is entirely possible to read the biological effect evidence as compelling as further attenuation of field intensities than established by the ultimate order. To that end, the PSC might have mandated the construction of a 345kV line within the same 350-foot right-of-way. However, by June 1978, a 765kV line had already been built. Indeed by June 30, 1976, the majority had rejected the 345kV option altogether, asserting that for technical reasons, one would need four or five such lines to carry the same load as a 765kV facility. The land impacts alone of that many structures militated against their choice.

What this assertion ignored was that the Hydro-Quebec contract called for 800mw of power, easily handled on a double-circuit 345kV line. A 765kV was required only if eventual loading reached 4,000mw. Yet there was no evidence that anything near that much power could be purchased from Canada. Rather as a quiet undercurrent to the proceeding was the idea that PASNY wanted so large a design capacity because it planned eventually to construct nuclear plants to intertie with the system.[48]

Ignoring the propriety of this long-term goal, never explicitly articulated by the parties, there was no reason to believe such plants could be sited in New York in the future. The economic and environmental arguments adduced by the commission on the question of hydropower would rationally have sustained certification of a 345kV line had the issues been honestly presented and had the commission not acceded to PASNY's purchases.

It has been suggested that, notwithstanding the abstract merit of the 345kV choice, it was never available to the commission because PASNY's ability to determine its own need extended to the size of the facility. But the matter is not as clear-cut as PASNY maintained; the fact that the PSC might have had to litigate the question should not have precluded its contemplation.

Moreover, assuming that the need issue did prevent the substitution of a 345kV line for one of 765kV, the PSC might still have achieved a further diminution of field effect and built in a further safety factor by extending the 765kV right-of-way an additional 250 to 550 feet from the center line (a total of 600 to 900 feet), as several intervenors argued.[49] If there had been no preliminary certification, route segment preparation, and early construction, the cost of such an extension would have been manageable; after the fact, according to PASNY, the economic consequences were too "substantial."[50]

The same can be said about conductor clearances. Were the PSC to have enunciated clearly its absolute disapproval when it learned that PASNY intended to make steel purchases, it might have prevented those purchases. If not, it would have preserved its unqualified right to ignore the purchases in making a final decision. A reading of the evidence on induced shock suggests that that danger was more possible than the probable worst-case postulated. The very shaping of the analysis was distorted by the apprehension of the existence of the purchases. Without that fact, exploration of the relative efficacies of mitigation efforts could have occurred in a way that raised no questions about the propriety of the approach taken to the issue. And had the PSC chosen to demand that PASNY purchase new goods, it would almost certainly have been sustained in court.

I became a member of the PSC less than a month before the final determination was handed down. I joined in that determination, for by then it appeared irreversible. I have asked myself since whether I would have seized the opportunities described had I had been part of the case from the beginning. To answer the question requires an analysis of why the commission majority acted as it did.

## Why the Commission Did What It Did

It is impossible to understand the commission's response to the application and its intermediate orders on that application without locating the matter in time. PASNY began its negotiations with Hydro-Quebec and its planning of the 765kV line just as New York started to feel the impact of the 1973 oil embargo. Con Ed's rates soared, and downstate consumers loudly demanded that the PSC take affirmative initiatives to reverse the prospect of what seemed to be unending rate increases.

Through the mid-1970s, substantial political pressure from municipal and state-elected officials built to move that inexpensive hydro as quickly as possible into the city's territory.[51] Environmental concerns, which some see as imposing economic costs on society, may suffer when competing with projects with quantifiable and immediate benefits. And when the risks are as unsusceptible to precise evaluation as they are in this instance, environmental caution becomes even less defensible a stance for an agency whose actions are open to constant review and criticism by the citizens whom it is charged to serve.

Finally, the PSC was dealing with another public entity. There is no doubt that were a private company to have done what PASNY did, particularly buying equipment that undercut the PSC's fundamental decisional freedom, the matter would have been treated entirely differently. First, the PSC could have expressed its righteous institutional rage at the hubris of the regulated utility. Second, it would have been comfortable assigning the loss associated with switching to alternative equipment to shareholders, since they should rightly bear the burden when their firm does business improperly. Here, however, there existed a substantially unregulated authority that lay claim to an expression of the will of the people as articulated by its legislature. Moreover, with such public entities, there is no way to distribute risk between rate payers and equity holders. The costs associated with scrapping equipment already purchased and buying replacements would have been passed on directly to customers: the city and its transportation units. Since the justification for the whole project was reduction in electric rates to those customers, loading onto them the responsibility of paying off additional capital investment would have violated the principle and conceivably vitiated the purpose.

Because of the time, the political pressure, and the nature of the utility with which it was concerned, the PSC ended, unwisely in my view, by circumscribing and perhaps undermining its own authority. Clearly it had the right to resist efforts to shortcut the health and safety case and to ignore unauthorized actions by PASNY. Perhaps because of the nature of the questions that the health and safety case raised, it had a responsibility to act with more procedural and substantive integrity. It is possible that the environmentally sound decision would have been a 345kV line or a 765kV line with higher conductor-to-ground clearances and a broader right-of-way. Imposing these as operating conditions after the full case was concluded would have strengthened the PSC's position in any subsequent court proceeding. By adhering to due-process requirements, the PSC would have proved unwilling not to be pressured by another agency into ignoring a clear statutory mandate established by the legislature.[52] Commission firmness might have reminded PASNY that it is not above the law.

## The Power Authority: Actions and Explanations

I have not yet discussed the general behavior of PASNY throughout these proceedings, except to note that it disregarded the PSC's February 1976 order to certify that it would build 345kV lines if so required and that it made material purchases specific to a 765kV line prior to PSC approval of that voltage level.

There was considerable public outcry against the line in the upstate communities through which it passed. People understandably complained that they were bearing all of the risks associated with the line and sharing none of the benefits. The PSC at least addressed these concerns, reminding the complainants that for years they had had the advantage of PASNY's cheap power when their downstate neighbors were not so served; that the state was indeed a single entity; and that the economic health of New York City had important consequences for the economic health and vitality of rural New York.[53]

PASNY, on the other hand, hardly deigned to offer explanations or justifications. Indeed at one point, when farmers took direct action by protesting nonviolently at the site of the line, PASNY officials had them arrested. Such behavior is not atypical of public authorities and raises very important questions about the level of confidence we should accord notions that government will do better by us than private enterprise.

PASNY's too-frequent arrogance and insensitivity can be explained in several ways. For investor-owned utilities, the theoretical reward is profits. Since profits are not an element of the public power calculation, something must replace them as a motive for action. What that often appears to be is the desire of those who run authorities to build monuments to themselves. Robert Moses immediately comes to mind. And James Fitzpatrick, who chaired PASNY for some years, presided over it when it constructed its first nuclear plant, the Fitzpatrick Nuclear Station.

Second, there exists for public authorities an urgent competition with private firms and a need to prove the truth of the yardstick theory. The pressure on the authority is to accomplish its goals at the lowest direct cost. There is a seemingly consistent tendency to ignore the price of externalities, such as the environment, and to be resistant to any notion that certain kinds of protection, which will clearly cost extra money, should be part of it.[54]

Vindication of the impulse to build monuments and demonstrate one's frugality requires avoiding regulation by other agencies. In some cases, the authority is aided in this goal by a legislature that may exempt it from showing need for transmission or generation siting and from having to hold evidentiary hearings before rates can be changed.[55] The authorities shore up their claim of independence by drafting bond indentures that specifically bind them to avoid activities that might involve economic losses (witness the Port Authority covenants against investing in mass transit) or are useful to advert to when the authority wishes another agency's regulations waived.[56]

Although it is probably true that all authorities share the desire to build monuments, to prove they can deliver more cheaply, and to establish autonomy, these factors do not entirely explain PASNY's continued resistance to recognizing that potential health and safety effects should be confronted, understood, and, if not benign, prevented.[57] Here we are dealing with a psychological phenomenon that may very well operate at all levels of government. There is an identity on the part of governmental entities with the public interest. If one is, in essence, the people one represents, it is almost impossible to acknowledge that what one does might cause harm, particularly if the project is undertaken with the best of conscious motives. Rather than admit that, notwithstanding these motives, damage might follow, one denies the very idea.

It is conceivable that in this case the PSC and PASNY will work out a settlement whereby PASNY voluntarily will offer some sum toward a research project.[58] This action will not be inconsistent with my theory, for there is a substantial psychological difference in being told by someone else that you might hurt your constituency and deciding, on your own motion, to exercise special, even excessive, caution in its behalf.

## Last Reflections

This 765kV line application represents a special instance of a general case. The questions it raises are, unfortunately, not confronted by Commoner when he envisions the solar future. Commoner's thesis presupposes an easily achievable identity between environmental and energy goals. Here such identity is missing. And here the presence of political and psychological pressures, common to many other kinds of governmental decisions, complicates the clarification of even the conflicting goals.

Certainly a partial solution might be found in administrative and legislative reform. For example, PASNY should not determine its own need.[59] Second, it would be preferable for an agency other than the PSC to make the initial health and safety findings. To develop the clearest evidentiary record, there ought to be an entirely undivided emphasis on public protection. The PSC necessarily weighs equities since the commission is also concerned about restraining price increases. Finally, in this context, the shape of Article VII creates difficulties because it fails to grasp directly the real possibility that economic considerations are probably not coextensive with environmental ones.

It would be disingenuous to assert that environmental-energy conflicts can be resolved by amending the administrative framework of the inquiry, for the difficulties are implicit in the very choices the application compels. It was never really a matter of whether or not to build the 765kV line. Rather the state ultimately faced deciding among unpalatable alternatives: between a 765kV or a 345kV system, both potentially hazardous; between an extra-high voltage line transmitting renewable hydropower or an 800mw nuclear plant to meet load requirements. Additionally our ability to assess the costs and benefits of any of these approaches was compromised by the lack of sophistication we have in quantifying the economic value of ecological, health and safety issues. Currently it is clear that the equation must tip in favor of the concrete price diminution expected from construction because the environmental impacts are more speculative, more remote, and politically less palatable to acknowledge.

There is one last important obstacle to the best resolution of the kind of energy-environmental problems that the 765kV case poses. Unhappily planning at state and federal level is in the hands of people who do not, for the most part, intend to make it a lifetime activity. Therefore, these decisions are made in the short run. Of course, some civil servants work for an agency for many years, but the directors of those agencies, the affirmative policy makers, change as the political complexion of government changes or as their ambitions develop. Environmental issues, indeed even energy issues, do not resolve themselves in the short term.[60]

So even when all participants are well intentioned, when legislatures have made authorities less insulated and more responsive, we are still likely, in projecting the future, to be too cavalier about consequences, too impatient with the uncertainties of our present knowledge, too precipitous in our reach for answers, too careless of environmental values that have yet to be well articulated. I do not suggest that we are consigned to inaction, but I think the case I have examined stands for the proposition that flexibility must be fought for, that any solutions are only proximate, and that no policy blueprint should rely on the assumption of inherent goodness or right thinking in a particular entity because it is ostensibly under public control.

**Notes**

1. This kind of approach resolves conflicts by creating situations in which the need for energy and the requirements of environmental health and safety are compatible goals.

2. Barry Commoner, "Reflections: The Solar Transition I and II," *New Yorker,* April 23, 1979, pp. 53-98, and April 30, 1979, pp. 46-93.

3. Commoner, "Reflections: Solar Transition II."

4. Long Island Lighting Co., Consolidated Edison Company, Orange and Rockland Utilities, Central Hudson Gas and Electric, Rochester Gas and Electric, New York State Electric and Gas, and Niagara Mohawk Power Corporation.

5. The governing statutes are the *Public Service Law* 47 (McKinney Laws, 1955) and the Transportation Corporations Law 62 (McKinney Laws, 1943).

6. Or 100-125kV extending ten miles or more: *Public Service Law,* S120(2).

7. The governing state statute is Public Authorities Law, Art. 5, title I, sec. 1000 et seq. Because PASNY has developed properties on water bodies and has tax-exempt status, it is also regulated by the federal government: Public Law 85-159, 71 Stat. 401; and the Internal Revenue Act, sec. 103(b) 4(E), et seq. (In practice, the chair is actually selected by the governor of the state and then the board votes to confirm.)

8. Public Law 85-159, 71 Stat. 401.

9. The terms of these convenants are negotiated by PASNY and bondholders in the indenture instrument. They bind both parties for the term of the instrument and are not constitutionally alterable by any act of the legislature.

10. Public Service Law (hereafter PSL), S126 (g) and Public Authorities Law (hereafter PAL), S1014. Two months after this case was decided, the Legislature amended PSL, S124, to require PASNY to prove need in the same way as any other applicant must. L. 1978, c.760, S1, eff. August 7, 1978.

11. This is the power agency of Canada's Quebec province.

12. At about the same time, though projecting much later construction, Rochester Gas and Electric and Niagara Mohawk Power Corporation filed to build a 765kV line to connect with a proposed nuclear plant. The application was withdrawn shortly after the final decision in the PASNY case but subsequently has been refiled.

13. PAL, S1001, 1001(a)

14. The Public Service Department, the administrative arm of the Public Service Commission, has a unit comprised of hearing officers, called administrative law judges. While the chair of the commission does appoint them and sets

their salaries and terms of their employment, they are otherwise independent of any direction from and oversight by the commission. Indeed they do not have ongoing or casual discussions with commissioners; public service commissioners' contact with them occurs only at public statement hearings, in cases where there are joint hearing proceedings and through their recommended decisions.

15. PSL, S122, lists the parties of right.

16. See PSL, S66(12), for rules on electric rate cases that are the same as those that obtain for gas, telephone, steam, and water tariff filings seeking to change the price of service.

17. There were also questions about the generation of ozone by the lines and the effects of the lines on pacemakers. In the first case, the danger was thought insubstantial; in the second, ultimately, an educational booklet was prepared and information about routing of the line was sent to the state's cardiologists.

18. Ordinarily the record is seen only after a recommended decision is written and briefs responding to it are filed. However, there is no rule that precludes the commission from receiving transcripts of the proceeding as they are completed and even against ordering parties to brief questions directly to the commission, bypassing the judge's recommended decision altogether.

19. Opinion 76-2, case 26529, Opinion and Order Granting Partial Certificate of Environmental Compatibility and Public Need, issued February 6, 1976.

20. Ibid., pp. 4, 5.

21. Ibid., pp. 5, 6. Essentially the commission's perception of the case at this point was that some line needed to be built to bring the power down, and it could decide later whether 345 or 765 was proper design kilovoltage.

22. Ibid., p. 6.

23. Letter of Chairman Fitzpatrick, March 10, 1976, cited by Order Amending and Clarifying Opinion, case 26529, April 1, 1976, p. 2.

24. Ibid., p. 1.

25. Opinion 76-2, p. 16.

26. Opinion 76-12, case 26529, Opinion and Order Authorizing Erection of Support Structures and Conductors, issued June 30, 1976.

27. Ibid., p. 4.

28. This is a motion common in civil practice.

29. Actually two days of cross-examination of a witness remained and, during that cross-examination, questions concerning the possibility of cancer were first raised. On rebuttal, in addition, some important issues were fleshed out in a manner that added considerably to their weight. However, the commission remained convinced that no fundamentally significant testimony had been ignored when it declared that it had considered the worst case.

30. Opinion 76-12, p. 5.

31. Some laboratory experiments with rats and mice in which they were exposed to fields similar to those at issue appeared to show that the animals were affected with respect to growth, (both cellular and of the total organism) and to the functioning of the central nervous and cardiovascular systems. One witness said that the condition of animals after thirty days' exposure was consistent with "chronic exposure to an environmental stresser." Questions were raised, however, about the reliability and applicability of the studies, and they were counterposed to others demonstrating no such effects. For an abstract of the studies showing effects that ultimately the commission relied on, see appendix A, pp. 1-6, Opinion 78-13, issued June 19, 1978.

32. Opinion 76-12, pp. 13-15.

33. Ibid., pp. 17-19.

34. The dissenting commissioner, Harold A. Jerry, Jr., pointed out that the fact that the Soviets strung lines lower than their studies recommended only "proves that the Russians are as reckless as we are." Opinion 76-12, Dissenting Opinion, p. 7.

35. Ibid., pp. 5, 6.

36. Opinion 76-12, pp. 20-32. Technically one can ground movable objects by using grounding chains. Therefore grounding could have been accomplished as far as concerned farm equipment regularly driven under the lines. However, one could not have predicted and anticipated every vehicle that might, either frequently or rarely, pass beneath the conductors and wires.

37. Ibid., pp. 6, 7.

38. Ibid., p. 7.

39. Ibid., p. 9. It was generally the contention of the commission majority that except for the level of audible noise produced in foul weather, 765kV lines were not qualitatively different from the overhead transmission and distribution facilities in use for decades. (Ibid., p. 5). The dissent said that this argument was misleading, for 765kV lines are quantitatively different; the electric field, for example, is much wider than the same strength field for a 345kV line. Ibid., Dissent, p. 15.

40. See Order Granting Further Partial Certificate of Environmental Compatibility and Public Need, issued December 29, 1976; Order Granting Partial Certificate of Environmental Compatibility and Public Need for Certain Route Segments and Denying Motion for Certification of Other Route Segments, issued June 21, 1977; and Order Granting Certificate of Environmental Compatibility and Public Need for Remaining Route Segments, issued January 12, 1978.

41. Opinion 78-13, Cases 26529, 26559, Opinion and Order Determining Health and Safety Issues, Imposing Operating Conditions and Authorizing, in Case 26529, Operating Pursuant to those Conditions, issued June 19, 1978. Case 26559 concerned the application of Rochester Gas and Electric and Niagara Mohawk Power Corporation to build a line. No decision on the

specifics of that line was made here, but the health and safety determination was binding on both companies. Also, although the record was regularly certified to the commission, a recommended decision nonetheless was written.

42. Ibid., p. 21.

43. Opinion 76-12, p. 5.

44. According to a rule adopted by the commission in the early 1970s, 2 percent of the gross cost of a line should be contributed by the applicant to develop compatible recreational uses of the right-of-way. The commission, in Opinion 78-13 argued that since recreational uses would be forbidden, this 2 percent, which would otherwise have to be spent on such projects, should instead be devoted to research.

45. Atwell et al. v. Power Authority; Upset, Inc. v. PSC; In the Matter of Power Authority of the State of New York v. PSC, Dockets 34696, 34586, 34489 (S.Ct. App. Div., 3d Dept., filed April 1979).

46. PASNY and PSC have finally negotiated a settlement. PASNY among other things has agreed to spend $50,000 towards the development of a research design. Then, after all parties accept the design Con Ed and PASNY together will contribute $5,000,000 to the research itself. However, this sum will not be sufficient to cover a comprehensive study; if therefore, other utilities do not supplement this sum, the settlement may be meaningless.

47. Opinion 76-12, Dissenting Opinion, p. 5.

48. Actually in ibid., p. 33, the commission does say that a line of the size contemplated is needed to allow "increased transfer capability for diversity power from Canada to meet summer loads in New York and to provide a more effective tie between the New York interconnected system and PASNY's existing upstate generating stations as well as any future generation that may be constructed in the area". However, there is no record support for the last phrase of the sentence.

49. See discussion in Opinion 78-13, pp. 13-14.

50. Cases 26529, 26559: Brief on Exceptions of PASNY, February 22, 1978, pp. 11-12.

51. I was a state senator during this period and remember the concern my colleagues and I shared about high Con Ed rates. I was also aware of the urgency with which Governor Hugh Carey viewed the line's completion.

52. By this, I mean the vesting in the commission of exclusive jurisdiction over siting matters. The fact that until 1978 PASNY could conclusively determine need does not vitiate the significance of the legislature's grant of authority, since on the numerous other findings necessary before a certificate could issue, the PSC's decision, not PASNY's, was controlling.

53. Opinion 78-13, p. 5.

54. The TVA, for example, has been notorious for its desire, until quite recently, to escape environmental regulation. PASNY has not diverged from a similar path.

55. The law has been changed, not only for transmission but for generation siting as well. The legislature did so in response to the behavior of PASNY in this and similar cases.

56. The Port Authority of New York and New Jersey bound itself until well after the year 2000 not to invest in projects like mass transit that might not be profit making. When the legislature attempted to repeal the convenants, over PASNY's vehement objection, so that some of the funds it had from other lucrative projects could help sustain the subways in the city, bondholders sued and succeeded in defeating the legislative intention.

57. Throughout the case, PASNY dismissed or denigrated any suggestion that potential harm might flow from the lines. In contravention of PSC orders, it began energizing the line at the same time it was insisting that no study on biological effects be undertaken. Indeed it was the direction that a study be done along with the rule on audible noise that precipitated the court suit. Clearly from their prior statements, PASNY could have lived with every other condition.

58. PASNY and the PSC have finally reached a negotiated settlement. The parties agreed to terminate their litigation, and PASNY agreed to mitigate noise impacts (although it will be arbiter of any disputes). PASNY also committed itself to contribute $50,000 in start-up costs for a health and safety research study and thereafter to contribute, with Con Ed, $5 million to the research. However, PASNY's chair will be one of three members (along with the chair of the PSC and the chair of the Department of Health) who chose the scientific panel to create the study outline, and PASNY will participate in monitoring its development. The original PSC order provided that the PSC would administer and monitor the research effort so its independence would be guaranteed. Consultant scientists have indicated that unless at least $13 million is forthcoming, no meaningful results can be achieved. The PSC-PASNY agreement thus contemplates but cannot guarantee voluntary contributions from other utilities.

59. Two months after the case was decided, the legislature amended the Public Service Law, Section 124, requiring PASNY to prove need in an identical manner as any other applicant. L. 1978, C.760, Sl, eff. August 7, 1978. It seemed clear from debate on the measure that the case provided impetus for the change.

60. Two-, four- and six-year time periods, the ones within which policy makers generally operate, are too limited to provide the care and continuity that planning needs.

# The Role of Local Government in Energy Policy

*David Harrison, Jr.,* and
*Michael H. Shapiro*

Energy issues have dominated many policy debates, since the 1973 Arab oil embargo dramatized the country's vulnerability to suppliers. The accident at the Three Mile Island nuclear plant added to the nation's concern for obtaining adequate energy supplies that are safe and reliable. Most federal policy discussions focus on solutions to the energy problem: measures to reduce consumption, increase supplies (particularly domestic supplies), and deal fairly with citizens in circumstances where serious imbalances between demand and supply occur. State officials have also been active in devising energy policies or recommending changes in federal policies, actions that underscore the differing vulnerability of regions of the country to energy shortfalls.

Consideration of the role of local governments in dealing with energy issues has become an important focus for policy development. Local governments can affect energy supply or regulatory policies—for example, through land use and permitting decisions that influence the siting of power plants and transmission lines. They may also play a significant role in implementing energy allocation strategies—such as California's gasoline-rationing plan—designed by higher levels of government. In general, however, we expect the primary role of local governments in energy policy to be to reduce the demand for energy in two major ways: by reducing their own consumption in operating public buildings and by encouraging private citizens to reduce consumption.

## Rationale for a Local Role

The arguments for and against a major role for local government in energy management and policy revolve around two basic questions. First, is it in the interest of local governments to play such a role? Second, is such a role appropriate given the characteristics of local governments?

### *Incentives for Local Government Involvement*

One argument for local involvement is based on political interests. To the extent that energy is perceived as an important issue in the community, local politicians

We wish to thank Larry Reilly for his assistance with this chapter.

have an incentive to assume strong leadership. It is possible, however, to gain the perceived political advantages associated with the energy issue without undertaking programs of substantial impact. Political incentives, though very real in many instances, do not ensure meaningful local government action.

A second line of argument relates to the incentives for maintaining the local tax base. As energy becomes increasingly expensive, communities with a large stock of energy-inefficient buildings and an infrastructure allowing inefficient transportation patterns would find themselves with a diminishing tax base. It is not clear how valid this argument is. Although housing values are likely to reflect differences in energy costs, communities nonetheless can adjust their tax rates to maintain desired service levels. The major effect of capitalized energy costs in the short run (assuming taxes are based on market value) therefore would be to redistribute tax burdens from energy-inefficient to energy-efficient property owners. The differentials might increase in the long run if neighborhoods with energy-inefficient housing units declined, although this scenario is highly speculative. In general, one would expect tax-base effects to be a relatively minor rationale for local government action.

Finally it can be argued that, at a minimum, the pressure of energy costs should lead local governments to conserve energy in their own buildings. By publicizing their own successes, local governments could encourage similar efforts on the part of residents and businesses. Since energy costs are still relatively small items in most local budgets, however, it may not be difficult to pass along the extra costs of inefficient energy management in the form of increased taxes.

### *Appropriateness of Local Government Involvement*

Perhaps the strongest argument in favor of a local government role in energy policy is historical precedent. Policies that could heavily influence energy use, such as building, zoning, and subdivision codes, as well as transportation investments, traditionally have been local responsibilities. For example, most municipalities have inspection departments that regulate building construction; regulating the heating and cooling of these structures is a logical extension of these activities. Zoning and subdivision ordinances affect the type of buildings constructed, their orientation, and the overall land-use characteristics of a community. With few exceptions these activities are exclusively local functions.

A second argument for a strong local role is that energy issues vary across different types of communities. Even within regions sharing a relatively similar climate, some localities may be growing slowly and have a large portion of their housing stock in old multifamily dwellings; other communities may be experiencing rapid growth, primarily in new single-family units. These local conditions mean that different energy-conservation strategies are appropriate. Local governments should be in the best position to determine strategies for their particular circumstances.

To illustrate this point, table 8-1 lists selected housing characteristics for eleven of the nation's largest Standard Metropolitan Statistical Areas (SMSAs) with a similar climate range of 4,000-6,000 heating degree days. The housing characteristics reflect features that can influence the effectiveness of government conservation strategies. For example, since rental units in large multifamily structures are less likely to have individual meters than are units in smaller structures, price-oriented policies will be less successful in areas like New York City where large apartment buildings comprise much of the housing stock. Programs that provide incentives to insulate or undertake other conservation measures will have the greatest impact where a large share of the stock is owner occupied. Moreover, the structural characteristics that affect a building's thermal inefficiency and ease of retrofit are likely to be related to the year of construction. Housing stock characteristics vary widely both among SMSAs and among communities within any single SMSA.

While energy conservation planning is closely related to the traditional responsibilities of local governments, most communities would require additional technical staff and associated resources to be effective. In addition, the implementation of conservation strategies may require more resources than

**Table 8-1**
**Selected Characteristics of Eleven SMSAs, 1970**

| *SMSA* | *Annual Heating-Degree Days* | *Percent Renter-Occupied Dwelling Units in Structures with More Than 20 Units* | *Percent Owner-Occupied Dwelling Units* | *Percent Units Built Pre-1940* |
|---|---|---|---|---|
| Baltimore | 4,654 | 4.0 | 58.2 | 39.5 |
| Boston | 5,634 | 3.9 | 52.6 | 63.9 |
| Cincinnati | 4,410 | 4.8 | 61.0 | 45.8 |
| Columbus | 5,660 | 4.2 | 59.0 | 34.1 |
| Indianapolis | 5,699 | 4.2 | 65.4 | 39.0 |
| Kansas City | 4,711 | 4.8 | 65.6 | 37.8 |
| New York City | 4,811 | 34.8 | 36.8 | 53.8 |
| Philadelphia | 5,144 | 7.2 | 68.6 | 51.4 |
| Seattle | 5,145 | 9.3 | 64.9 | 30.1 |
| St. Louis | 4,900 | 3.6 | 64.7 | 41.6 |
| Washington, D.C. | 4,224 | 16.6 | 46.0 | 21.7 |
| Range | | 3.6-34.8 | 36.8-68.6 | 21.7-63.9 |
| Range (excluding New York City) | | 3.6-16.6 | 46.0-68.6 | 21.7-63.9 |
| U.S. average | | 6.3 | 62.9 | 43.4 |

Source: U.S. Census Bureau, *Census of Housing: 1970, General Housing Characteristics* (Washington, D.C.: Government Printing Office, 1972).

local governments are willing to commit. Since subsidies probably come from state and, federal sources, the scope for local initiative is inevitably reduced.

Although a case can be made for local involvement in energy conservation issues, both the incentives for action and the ability to follow through are somewhat limited. Even when local involvement is strongly justified, action at other levels of government may restrict the scope of local initiative. Nevertheless, the variation in local characteristics suggests potentially useful roles for local governments. These governments are in the best position to respond flexibly to the needs and priorities of their communities.

## Conservation in the Local Public Sector

### *State and Local Government Energy Consumption*

Energy consumption figures for city and county governments are not available, but Hoch's recent study of energy consumption provides estimates for a combined state and local government sector.[1] The data presented in table 8-2 also include energy use in terms of primary fuel consumption, obtained from Hoch's data by distributing electric use among primary fuel sources and adjusting for conversion efficiencies. Overall state and local government energy use accounts for about 3.7 percent of the nation's primary use, or about 2.5 quads ($10^{15}$ Btus) per year.

**Table 8-2**
**Energy Use by State and Local Governments**

| | *Petroleum Products* | *Natural Gas* | *Electricity* | *Coal* | *Hydro* | *Nuclear* | *Total* |
|---|---|---|---|---|---|---|---|
| End use | | | | | | | |
| $10^{12}$ Btu | 544.6 | 680.3 | 383.6 | | | | 1,608.5 |
| Percent of nation | 1.9 | 3.0 | 7.1 | | | | 2.2 |
| Primary energy use[a] | | | | | | | |
| $10^{12}$ Btu | 763.8 | 974.4 | NA | 553.1 | 183.9 | 36.4 | 2,511.7 |
| Percent of nation | 2.6 | 4.2 | NA | 4.5 | 7.1 | 7.1 | 3.7 |

Source: Irving Hoch, *Energy Use in the United States by State and Region* (Baltimore, Md.: Resources for the Future, 1977).

[a]Estimated from Hoch's end-use data by assuming an overall efficiency of 30 percent for converting primary energy to electricity and using the national average fuel mix for electric generation.

NA: Not Applicable

As table 8-3 demonstrates, natural gas is the largest single energy source for state and local governments, accounting for 42 percent of their consumption. Hoch estimates that approximately 62 percent of the energy consumed by this sector was used for space heating, 21 percent for interior lighting, 12 percent for transportation, 3 percent for street and highway lighting, and 2 percent for industrial processes.

Since building-related functions account for the great majority of state and local government energy use, a crude disaggregation between the two can be based on floor areas. Based on estimates by Ide et al. of floor space per employee, together with U.S. Census Bureau data on government employment, we conclude that local government energy consumption accounts for about 83 percent of total state and local government use, or about 3.1 percent of the nation's primary energy use.[2]

### *Potential Energy Savings*

Because most local government energy use is for space heating and cooling, major reductions in these categories must be achieved in order to cut total use substantially. Although comprehensive studies of potential energy savings in government buildings are not available, government structures are similar to commercial office buildings, which have received much more attention. The available information suggests that opportunities for substantial savings exist in both new government buildings and older structures. For example, modest, economically justifiable improvements in new-building energy consumption can increase efficiency by 69 percent for space heating and 36 percent for cooling, relative to 1970 buildings.[3] Measures such as better heating,

**Table 8-3**
**State and Local Government Energy Use, by Energy Source and Purpose**

| | *$10^{12}$ Btus* | *Percent of Total* |
|---|---|---|
| Petroleum products | 544.6 | 33.9 |
| Gasoline | 194.6 | 12.1 |
| Heating fuels | 350.0 | 21.8 |
| Natural gas | 680.3 | 42.3 |
| Industrial | 33.7 | 2.1 |
| Heating | 646.6 | 40.2 |
| Electricity | 383.6 | 23.9 |
| Building (lighting and air conditioning) | 342.0 | 21.3 |
| Street lighting | 41.6 | 2.6 |

Source: Derived from Hoch (1977).

ventilation, and air-conditioning (HVAC) system maintenance, changes in operating policies (including improved control systems), and insulation of pipes and ducts can also improve the heating efficiency of existing buildings by up to 60 percent.[4] Over the next decade, local governments should be able to improve the thermal efficiency of their buildings by up to 50 percent relative to 1970.

Energy used to light government buildings can be reduced by 33 percent over 1970 figures by appropriate reduction in the amount of lighting provided and better management.[5] In addition, local governments can conserve energy as more efficient vehicles replace existing fleets. Although performance requirements for police cars and emergency service vehicles limit these improvements, a suitable fleet mix should allow local governments to increase their transportation energy-use efficiency more or less in line with the rest of the nation.

### *Local Public Conservation Efforts*

Although many opportunities for energy-efficiency improvements exist, it is difficult to determine whether local governments are, in fact, achieving these savings. Governments may respond by acting as a model of energy efficiency because of the political visibility of the issue; by minimizing costs in an attempt to keep overall expenses down; or by making minimal efficiency improvements and passing energy costs on to taxpayers. Undoubtedly the response will vary from one locality to another. Although we have not examined these hypothetical responses in light of actual experience, we have investigated the energy use for a number of buildings, both government and private, for which data were readily available.[6]

Figures 8-1 and 8-2 indicate that the Boston Public Library substantially reduced its steam and electric energy consumption, primarily through simple and inexpensive conservation measures. The dramatic drop in steam consumption between 1973 and 1974 occurred after the library purchased a $30,000 air-conditioning unit for the rare-books collection; this investment made it possible to shut down the HVAC systems in the two main buildings for nine hours every night. Electricity consumption was also reduced by removing nonessential lighting and minimizing waste and by the nine-hour building shutdown. By 1977 steam use was less than a third of 1973 levels, and electricity use had fallen by more than 50 percent. These increases in efficiency were achieved without shortening library hours or removing lighting from public areas.

Figure 8-3 graphs the natural-gas consumption per heating degree day for a small sample of public and private buildings in Boston. Although the data are not appropriate for statistical analysis, we can make certain observations. All buildings for which energy data are available reduce gas consumption from 1973 to 1974. Private consumption has increased somewhat since 1974 but has

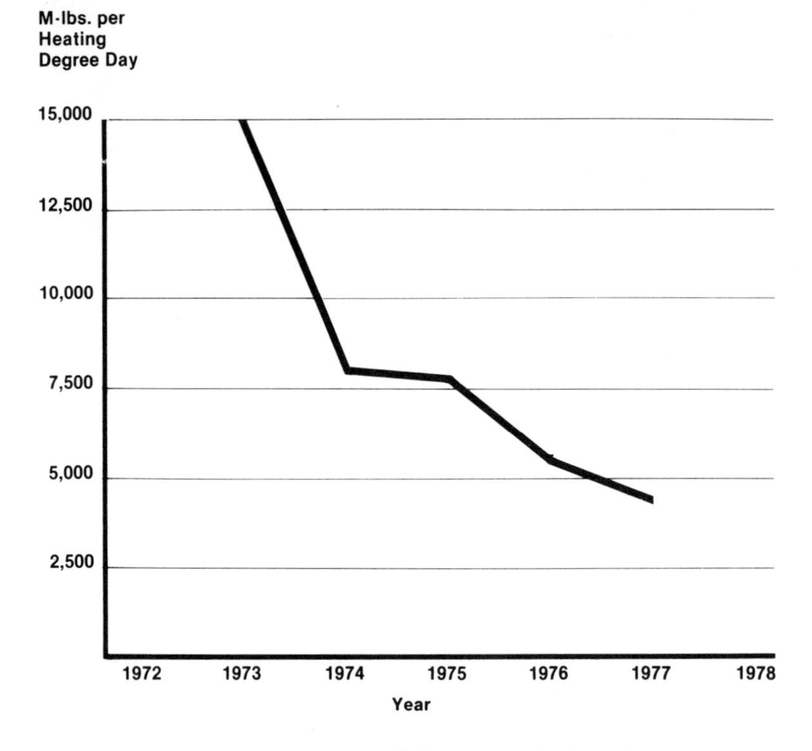


**Figure 8-1.** Steam Consumption of the Boston Public Library System, 1972-1978

not returned to pre-1973 levels. Although more variable than private use, public-building energy use has generally declined since 1976. We do not have sufficient data to explain the increase that occurred in 1976 in most sample public buildings.

## Local Government Promotion of Private Conservation Efforts

### *Savings in Residential Energy Use*

Residential energy consumption, estimated at $11.15(10^{15})$ Btus at point of use in 1974 or about $15.22(10^{15})$ Btus of primary energy, represents 19 and 21 percent of the respective national totals.[7] Since about 70 percent of this use is devoted to space heating and cooling, significant reductions in residential heating

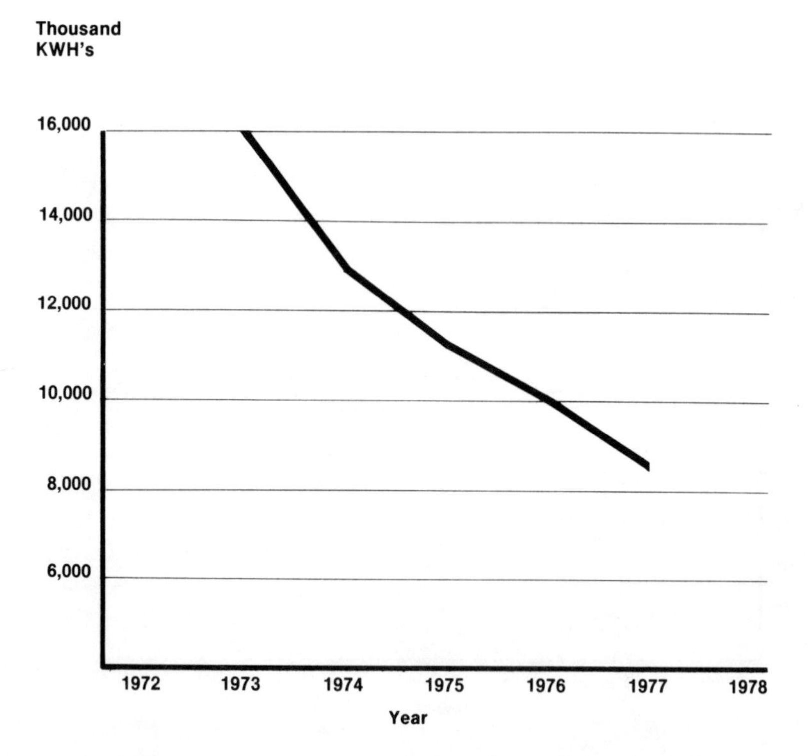


**Figure 8-2.** Electricity Consumption of the Boston Public Library System, 1972-1978

and cooling energy consumption would contribute substantially to national energy conservation. Moreover it appears that residential heating and cooling fuel-use patterns can be substantially modified, although estimates vary as to the amount of savings possible. A recent study, for example, concluded that the combined effects of stock mix, improved appliance efficiency, and improvements in the thermal performance of new and existing units would result in 15 percent less energy use in the year 2000 than under 1970 baseline conditions.[8] Moreover this analysis assumes very modest thermal improvements relative to what can be achieved in residential structures.

There is some uncertainty about how much residential energy use can be reduced through the ordinary workings of the price system or through government intervention. High prices will tend to induce energy conservation even in the absence of special governmental regulations concerning, for example, the insulation of buildings. Evaluations of government programs have sometimes

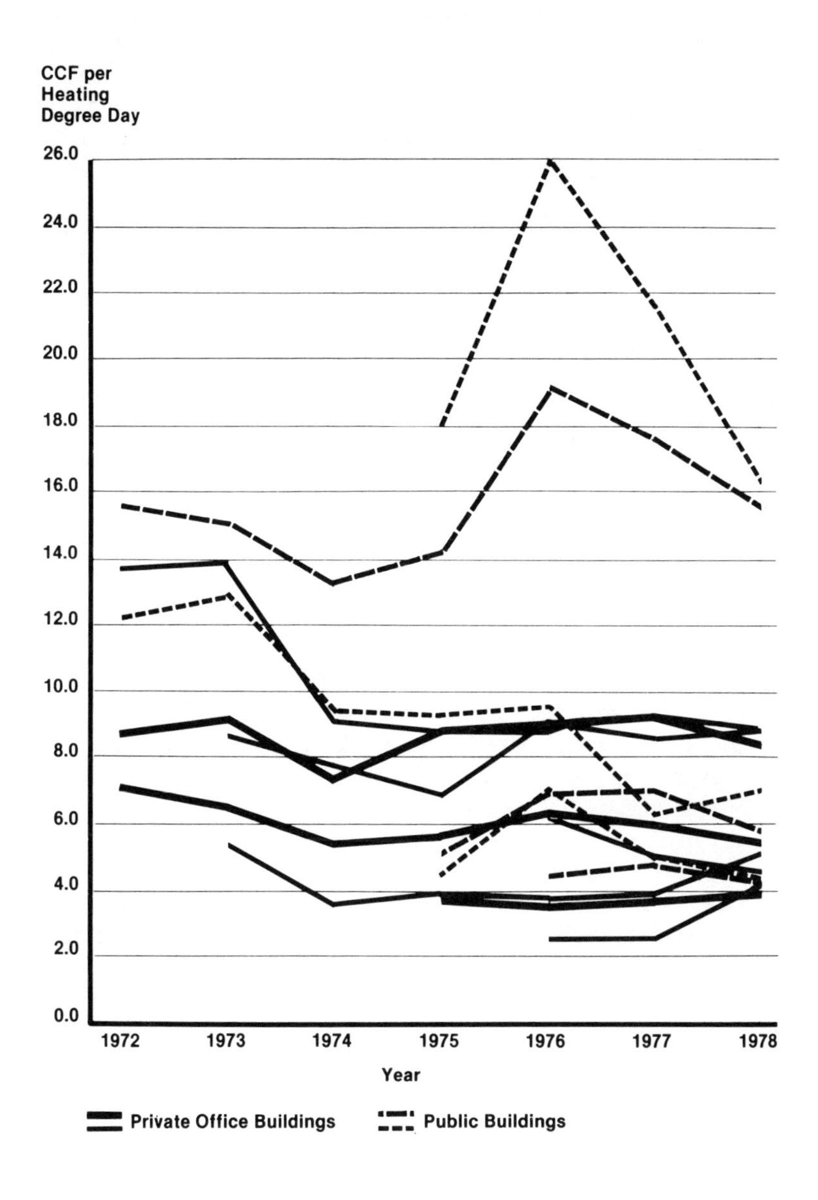


Source: Data from Boston Gas.

**Figure 8-3.** Natural Gas Consumption

obscured this distinction. For example, one evaluation of federal energy conservation efforts assumes that government regulations determine structural characteristics and appliance efficiencies, while price affects energy use for given structures and appliance types; the analysis may thus seriously overstate federal program impacts.[9] Nonetheless, some government action may be appropriate to overcome market distortions, accelerate processes that would occur more slowly, and deal with equity issues. At the local level, four general approaches have been suggested:

1. The use of building codes and related ordinances to promote conservation in new buildings.
2. The use of zoning policies and subdivision regulations to encourage structure types that minimize energy use.
3. The use of habitation codes to permit energy conservation in rental buildings.
4. The provision of information and technical assistance, such as energy audits, to increase public awareness of the economic benefits of conservation investments.

**Building Codes and Related Ordinances.** Numerous studies have indicated that residential structures can have considerably better thermal characteristics without drastic changes in construction. For example, the American Society of Heating, Refrigerating and Air-conditioning Engineers (ASHRAE) standard 90-75 suggests energy-efficiency levels for new building designs. Single-family units constructed to these standards would have 11 percent lower heating loads and multiple-family units would have 54 percent lower loads, compared to similar buildings designed in 1970.[10] The ASHRAE standards are not very strict, particulary for single-family dwellings and some studies suggest that savings on the order of 50 percent are economical on a life-cycle cost basis.[11] Existing buildings are also of concern. Since only about 1 percent of the housing stock turns over each year, close to half of the nation's residential units will still be of 1970 or earlier construction in the year 2000. Even for existing buildings, however, substantial improvements in efficiency are possible through additional insulation and other measures. Assuming that all units built after 1980 can be made 40 percent more efficient than 1970 units, all 1970 and earlier units are retrofit to be 20 percent more efficient, and units built between 1970 and 1980 achieve improvements halfway between these values, the average thermal efficiency of the housing stock would improve by 25 percent by 1985.[12]

Changes in building codes are one means of ensuring that high thermal performance standards are met. In the case of new buildings, however, federal and state actions have largely preempted local initiatives. The Energy Conservation Standards for New Buildings Act of 1976 (Title III of Public Law 94-385) requires that the Department of Energy and the Department of Housing and

Urban Development establish energy-performance standards for most new buildings. In the interim, state conservation programs, authorized by the Energy Conservation and Production Act of 1976 (PL 94-385), include energy codes for new buildings, most of which are based on ASHRAE 90-75.

It is generally assumed that existing local institutions will be largely responsible for implementing these standards. Performance-based codes, however, are substantially different from the component and practices standards traditionally imposed by local building-inspection departments and require a more-sophisticated understanding of the interrelationships among building components. Moreover there may be significant differences between the design thermal performance of a structure and its actual characteristics; relatively small imperfections in construction details, difficult to detect and not dangerous to the integrity of the building, can result in substantial heat loss.[13]

Federal regulations will not cover existing housing, leaving considerable discretion to local governments. Communities with large stocks of older, poorly insulated dwellings might consider requiring retrofit up to certain minimum degrees of performance, to be implemented perhaps at the time of sale. Availability of federal tax credits under the 1978 National Energy Act makes retrofit less expensive to the home owner. Localities may be able to develop easily enforceable criteria. For example, if the housing stock consists largely of single-family homes of similar style and age, a community should be able to specify retrofit packages that are easy to review and verify. Despite the potential savings that such a program could realize, problems do exist. Residents of one community may resist programs that place greater burdens on them than on residents of other communities. Action at the state level may therefore be necessary, although enforcement responsibility would fall primarily to local governments.

**Zoning and Subdivision Standards.** Traditionally local governments have been responsible for controlling the location and type of land uses within their boundaries, subject to constitutional limitations. It has been argued that these powers should be used to encourage types of structures and site designs that minimize energy use.[14] One study of residential structures in nine U.S. cities, representing different climate regimes, indicates that combined heating and cooling loads per square foot of living area in low-rise, multifamily units are from 27 to 80 percent lower than in single-family units.[15] Although these results must be qualified because typical construction practices for different types of units use different insulation practices, other studies have also confirmed the greater efficiency of attached and multifamily structures.[16] If single-family units were to comprise only 25 percent of new construction instead of 50 percent and attached and multifamily units were on average 30 percent more energy efficient than single-family detached, the overall efficiency of new housing would increase by about 10 percent.[17] But it is important to recognize that different conservation measures are not additive. For example, if federal performance standards substantially increase the thermal efficiency of all new residential structures, the additional savings from a change in housing mix will not be as great in absolute terms as in the absence of such standards.

Siting relative to wind patterns, screening vegetation, and orientation to the sun can also influence energy use. It is not possible to generalize about the potential savings from such measures, but one source suggests that they are significant though probably less than the savings possible from changes in structure type.[18] The greatest opportunities for energy conservation exist in developments where overall densities are relatively low—so there is a maximum flexibility for site layout—and where solar energy can be used economically.

If market forces generate a greater demand for attached and multifamily units, municipalities can facilitate developers' efforts by providing zoning and subdivision codes that accommodate mixed residential uses and by designing appropriate infrastructure. In addition, local governments can encourage market forces by providing incentives such as density bonuses for developers who propose desired housing types. Zoning policies that allow planned unit developments (PUDs) or cluster developments would give developers the flexibility to provide fuel-efficient structures and to improve site layout.

Although municipalities have the powers to undertake these actions, suburban communities traditionally are unreceptive to changes that encourage apartment or even townhouse construction.[19] Many issues—including fiscal impacts, exclusionary desires, and property-value considerations—influence these feelings, and it is unlikely that energy use could exert a major influence in this context. Nevertheless many communities have developed more-flexible zoning policies and have used PUD and cluster-zoning ordinances effectively. As experience with these measures broadens, it is likely that they will gain more acceptance, but it is not clear that energy considerations per se would substantially influence suburban development patterns.

**Habitation Codes.** Many local governments establish codes to control the management of rental property, including the setting of minimum temperatures during the heating season. In communities that have significant numbers of rental housing units with central heating sytems, changing the minimum allowable temperatures may seem an easy means of saving energy. Lower thermostat settings have been estimated to save energy at the rate of 2.5 to 4 percent per degree reduced.[20] Given the incentives for fuel saving, however, most building managers probably ignore codes that require high temperatures. Code reductions, therefore, would not necessarily translate into reductions in actual heating levels.

**Information.** Finally local governments may attempt to promote conservation by providing information and technical assistance through energy audits and similar programs. This strategy is attractive because it provides a high degree of visibility and relies on voluntary participation. As in the case of new-building-performance standards, however, federal legislation may have substantially reduced the opportunity for independent local action. The National Energy Conservation Policy Act of 1974 requires utilities and home-heating fuel suppliers to make energy audits available to their customers by 1980, thus obviating the need for local government efforts. If strong government-sponsored audit and assistance programs already exist, it might be appropriate for utilities to arrange cooperative undertakings with the communities.

There is still some uncertainty about how effective audit programs will be either in accelerating the rate of retrofit investment or in inducing investment that might not otherwise occur. Available studies suggest that a high percentage of home owners who receive audits undertake some, though not all, recommended measures. One follow-up survey of a Massachusetts program indicates that 60 percent of the home owners responding undertook some improvements, but only 30 percent of the measures recommended were actually undertaken.[21] Studies of this sort are plagued with difficulties: participants in the initial voluntary programs are more likely to be interested in conservation in the first place; and those responding to the follow-up surveys are also more likely to have taken action than those who did not. In addition, the surveys cannot measure long-term effects very well. Nevertheless, this evidence suggests that the information strategy can be effective. The availability of private alternatives and the federal mandate for action by utilities, however, may make major local government audit programs unnecessary.

### *Savings in Transportation Energy Use*

Highway transportation accounts for 21 percent of total U.S. energy use and about 44 percent of national petroleum use. Much of highway travel is passenger travel, dominated by the private automobile. Indeed many efforts to decrease U.S. dependence on foreign oil focus on curtailing gasoline use. National proposals such as closing stations on Sundays or increasing parking charges have been resisted, partly because transportation planning traditionally has been a local government function. The five major options to reduce petroleum use in urban passenger travel center on decreasing gasoline consumption by:

1. Increasing the fuel efficiency of the travel modes, particularly the automobile.
2. Reducing automobile use and increasing mass transit use.
3. Decreasing trip frequencies.

4. Reducing the amount of travel made under highly congested, fuel-wasting conditions.
5. Shortening average trip lengths.

**Improving Fuel Efficiency.** This strategy is least amenable to local control because of federal legislation. Under the Energy Policy and Conservation Act, Congress established increasingly strict new-car fuel-efficiency standards. By model year 1985, cars must meet a standard of 27.5 miles per gallon (mpg), more than a doubling of the 1974 fuel economy level of 13.3 mpg. The actual fuel efficiency may well exceed 27.5 mpg by 1985. After the entire fleet is converted to the fuel-efficient cars, energy consumption per mile of urban travel will be reduced by approximately half. Although there is some debate about the correspondence between test results and actual mileage, there is no doubt that the future automobile fleet will be much more energy efficient than present vehicles.

**Changing Mode Split.** Encouraging mass-transit use is perhaps the most common proposal to reduce gas consumption, partly because shifts from automobile to transit are considered desirable for other reasons. Recent studies, however, suggest that these proposals may be misguided, at least in terms of energy savings. The Congressional Budget Office (CBO) recently estimated energy use in British thermal units (Btus) per passenger mile for the automobile and a variety of alternative modes.[22] The data in table 8-4 distinguish three measures of energy use: operating or propulsion energy use; door-to-door energy use, including the energy characteristics of feeder operations, the roundaboutness of typical travel patterns, and the energy needed to build and maintain road, track and equipment; and program energy or the average savings for diverting a passenger mile to the mode. This last measure takes into account the modes from which each mode is expected to get its new passengers.

The CBO results indicate that programs designed to increase nonautomobile travel will save much less energy than many proponents of mass transit have suggested. The evidence is particularly damning for heavy rail systems; while requiring much less operating energy than the single-occupant automobile or the average automobile (1.4 persons), building new rail systems actually increases energy use when construction, travel to and from stations, and the roundaboutness of rail travel are considered. The most attractive options are van pool and car pool, but we question how much could be gained by local government support for those programs. The van pool is relevant only for a small segment of the travel market. Although car pools have a wider potential market, experience to date suggests that the influence of public (or spirited private) programs on the level of participation is quite small; incentives to car pool come primarily from higher gasoline prices rather than from government publicity or computer matching.

**Table 8-4**
**Middle Estimates for Various Measures of Energy Required by Urban Transportation Modes**

| *Mode* | *Operating Energy*[a] | *Modal Energy*[b] | *Program Energy*[c] |
|---|---|---|---|
| Single-occupant automobile | 11,000 | 14,220 | N/A |
| Average automobile | 7,860 | 10,160 | N/A |
| Car pool | 3,670 | 5,450 | 4,890 |
| Van pool | 1,560 | 2,420 | 7,720 |
| Dial-a-Ride | 9,690 | 17,230 | (12,350) |
| Heavy rail transit (old) | 2,540 | 3,990 | N/A |
| Heavy rail transit (new) | 3,570 | 6,580 | (980) |
| Commuter rail | 2,625 | 5,020 | 970 |
| Light rail transit | 3,750 | 5,060 | 30 |
| Bus | 2,610 | 3,070 | 3,590[d] |

Source: Congressional Budget Office, *Urban Transportation and Energy: The Potential Savings of Different Modes* (Washington, D.C.: Government Printing Office, 1977).

Note: All measures are expressed in Btus per passenger-mile.

[a]Propulsion only.
[b]All forms of energy, computed on a door-to-door basis, adjusted for roundabout journeys.
[c]Energy saved (lost) per passenger-mile of travel induced by new programs.
[d]For new express bus service. Regular urban bus service would show smaller savings.

Because they do not involve large access energy costs and because service can draw heavily from the automobile, express bus systems are predicted to generate the largest savings. Local governments, through their transit authorities, can thus promote conservation by encouraging express bus travel (for example, giving buses priority in traffic by means of special signaling or exclusive rights-of-way). Energy savings from adding express bus service, however, will be smaller than the CBO estimates, which are based on current automobile fuel efficiency of approximately 11.3 miles per gallon rather than the federally mandated standard of 27.5 mpg. In the short to medium run, when the automobile fleet is dominated by large gas-guzzlers, express bus service may thus provide substantial energy savings; as the automobile fleet becomes more fuel efficient, the gains will decline. Moreover the incentives for drivers to forsake their cars for an express bus will diminish as the per-mile cost of auto travel decreases. Although the more fuel-efficient automobiles are likely to provide less comfort and convenience than larger cars, they will still provide greater convenience than the bus.

**Decreasing Trip Frequencies.** Higher gasoline prices cause drivers to reduce the number of trips. Although work trips are probably not affected, employers may institute four-day work weeks as commuting becomes more expensive.

The frequency of nonwork trips may be reduced more substantially, although automobile fuel efficiency improvements will offset higher gas prices. Nevertheless, households are likely to make fewer shopping trips, personal visits, or pleasure trips as gasoline prices increase or supplies become less available.

One of the major proposals for reducing trip frequencies is to curtail weekend travel by closing gasoline stations on Sundays. Whatever the desirability of this strategy at the state or federal level, Sunday closings are not a viable local policy. One reason is that many local governments have small jurisdictions, which would make the program relatively ineffective; perhaps more importantly, the ban would be very unpopular among a local politician's constituency.

**Increasing Travel Speeds.** Fuel-consumption studies indicate that automobile miles per gallon depend on average travel speeds. Stop-and-start driving uses much more fuel than travel under relatively free-flow conditions. Researchers at General Motors have estimated that increasing the average speeds in the most-congested areas could result in average fuel savings of up to 2.5 percent.[23] These improvements might be achieved by widening roads and intersections, eliminating on-street parking, adding turning lanes, or restricting truck use. In addition, if all trips occurred at the posted speed limits, automobile fuel consumption could be reduced by an estimated 15 percent.

Although most local transportation planning efforts seek to increase travel speeds in congested urban areas, there are obstacles to transportation management practices and capacity improvements. In the large, older cities with the most serious congestion, narrow streets and high-density land uses make widening streets or adding turning lanes difficult and expensive. Enforcing on-street parking bans or bans on truck traffic is also difficult and may work only when combined with off-street parking projects, a costly proposal that might actually attract more traffic. Indeed the initial success of traffic management schemes may be partly offset by changes in travel patterns that they produce. If less-congested conditions generate more and longer trips, as many transportation analysts suggest, fuel consumption may not decrease significantly.

**Shortening Average Trip Lengths.** Reducing vehicle miles of travel (VMT) by decreasing the average length of a trip is attractive both because it unambiguously reduced gasoline consumption and because it seems particularly amenable to local action. Average trip lengths clearly depend upon residential and employment patterns, which local governments traditionally regulate. Several commentators have suggested that local governments can use zoning, subdivision regulation, infrastructure investment, and property tax policies to generate land-use patterns that economize on urban travel.[24] Evaluating the effectiveness of local efforts to conserve energy by changing land uses requires determining which patterns conserve energy, what policies shape urban land-use patterns, and which urban areas are likely to benefit from these policies.

A recent study by the Urban Institute summarizes the available evidence on the links between urban land-use patterns and energy use in passenger travel. The evidence consists of three types of studies: simulation studies of urban travel and energy use under different scenarios, a study of travel behavior in various neighborhoods in selected samples of metropolitan areas, and an econometric study of urban gasoline use. The results are both consistent with each other and with common sense: the gasoline-efficient city is small, compactly developed, has a large proportion of its population living in high-density neighborhoods, and has a relatively uniform distribution of jobs and population.[25] The Urban Institute estimates of average daily household VMT for Los Angeles, Youngstown, and the average area in their sample are presented in table 8-5. These figures suggest that channeling growth to centralized, high-density locations can save approximately 50 percent of VMT (and to a first approximation, energy use in travel) compared to fringe, low-density development.

The simulations and other studies of urban travel imply that the relative proximity of employment and population influences urban trip lengths. In other words, when people live close to their work places, average trip lengths are short, whether the jobs are in the suburbs or the center city. Indeed the simulation studies demonstrate that the combination of concentrated jobs and dispersed population generates the greatest travel demand and energy use. If households are dispersed, the best strategy is to encourage decentralization of employment. For the typical urban area with employment more concentrated than residences, energy savings would accrue either from high density residential growth close to the center or from employment growth in the suburbs. A sensible energy-conserving strategy would encourage both population and employment growth in suburban subcenters.

The local options for encouraging energy-efficient land-use patterns include zoning for higher-density development, zoning to allow intermingling of employment and residences, and using infrastructure investments (such as roads and

**Table 8-5**
**Household Automobile Travel Associated with Alternative Neighborhood Development Patterns**

| | *Household Daily Miles of Auto Travel by Neighborhood Type* | | |
|---|---|---|---|
| *Area* | *Inner High Density* | *Fringe High Density* | *Fringe Low Density* |
| Los Angeles | 49.7 | 73.5 | 101.9 |
| Average area | 24.8 | 35.9 | 52.7 |
| Youngstown | 14.7 | 20.2 | 29.6 |

Source: G.E. Peterson and W. Bateman, eds., *Urban Development Patterns* (Washington, D.C.: Urban Institute, forthcoming).

sewers) or permit systems to control the timing and location of development. Although it is not possible to provide complete quantitative evaluations of these and similar policies—or to determine their nonenergy implications—a few comments are in order. Traditional suburban zoning controls tend to discourage high-density residences and most employment, particularly manufacturing. The increasing importance of service jobs and clean industries (electronics firms and the like), the ability of multifamily developments to preserve a lower-density ambiance, and the fiscal advantages of higher-density development may, however, decrease suburban resistance to zoning changes of this sort. Planning the location and capacity of transportation and sanitation facilities can also have an effect on residential growth, although basic demographic, income, and price trends limit the impact of infrastructure investment.[26] Although there has been much recent discussion of efforts by towns such as Ramapo, New York, to use novel development permit systems to limit or guide suburban growth, their influence is questionable.[27] Most suburban developers already want to locate relatively close to existing developments so that employment and shopping areas are accessible.

Which urban areas stand to gain large gasoline savings from adopting these policies? The crucial determinant is the area's growth rate; residential capital is extremely durable, and most of the effect of land-use policies will be on new development. Thus quantitative estimates based on de novo simulations or urban area comparisons should be viewed as extreme upper bounds for the changes possible from implementing land-use policies to bring people and jobs closer together. Some studies suggest that the savings in urban travel and energy consumption could be as much as 50 percent, but this figure could be approached only in a rapidly growing area where existing development is a small fraction of the total. In older urban areas where growth is slow, the impact of land-use policies will be very small.

## Summary of Conclusions

We have considered the role of local governments in encouraging energy conservation by reducing both their own energy use and the residential and transportation energy use of their citizens. The overall conclusion, necessarily guarded because of limited information on many key factors, is that local governments will play a relatively small role in the country's overall energy conservation efforts.

### *Local Government Energy Uses*

The combined energy used to heat and cool local public buildings, operate government automobiles, light streets, and perform other functions is considerable; we estimated that these uses account for over 3 percent of total national primary energy use.

Local government officials are likely to respond to higher energy prices by purchasing energy-efficient capital (building and automobiles), by making existing capital more energy efficient, and by eliminating unnecessary energy use. Without more detailed data, however, it is impossible to determine how successful local governments have actually been in conserving energy and whether they could improve if given additional incentives.

### *Reducing Residential Energy Use*

Local government actions to reduce home energy use are likely to have modest effects for several reasons. Prices already provide substantial incentives for residents to modify their housing to reduce energy use, and the federal government has preempted the setting of energy-efficiency standards for new dwellings. Local governments may play an important role in enforcing federal standards, but they face major difficulties in designing inspection procedures and training inspectors to detect imperfections in construction.

Local governments may play a larger role in encouraging residents to retrofit their homes for energy efficiency. They might, for example, specify retrofit packages for single-family units that are easy to verify and could be required when a house was sold. In addition, local governments could provide information on energy-saving steps that home owners and apartment owners could take through energy audits and similar programs. Residents may, however, resist mandatory programs that place greater burdens on them than other communities (even though federal tax credits reduce the burden). In addition, the federal government substantially reduced the need for local information programs by requiring that utilities and home-heating fuel suppliers make energy audits available to their customers.

Local efforts to modify the mix of new housing toward multifamily units, which require less energy for heating and cooling, are not likely to be successful. While zoning and subdivision controls may influence the mix of units, particularly by allowing builders to construct multifamily units in cluster zones or planned unit developments, suburban governments are likely to base zoning decisions on nonenergy issues. Moreover federally mandated improvements in the energy efficiency of single-family units reduce the energy payoff of encouraging multifamily units.

### *Reducing Transportation Energy Use*

The largest savings in urban travel energy use will probably come from a combination of higher gasoline prices and federally mandated fuel-efficiency standards that by 1985 should make new cars approximately twice as efficient as the precontrol (1974) automobile. These improvements in automobile fuel efficiency lessen the energy savings from all other conservation measures, since all are designed to decrease automobile travel.

Of the major local options for encouraging energy conservation in passenger travel, constructing fixed-rail systems appears to be the worst. Indeed when account is taken of the energy used in construction and in getting to the rail stations, adding a rail system may actually increase rather than decrease energy use. While paratransit options (car pool and van pool) may offer substantial savings, local government policies are not likely to encourage their use significantly. Encouraging express bus service could decrease automobile use, although the market for such systems may be quite small given improvements in automobile fuel economy.

Policies to lower automobile VMT by reducing trip lengths may provide the greatest opportunities for decreasing fuel use. These measures include a variety of zoning, subdivision, infrastructure investment, and tax policies intended to bring jobs and population closer together.

Decreases in urban travel and energy use can also be achieved by influencing the patterns of growth. The increasing skepticism toward growth for a variety of reasons (air and water pollution, loss of open space, and diseconomies of scale in the provision of various public services) may cause large urban areas to curtail growth and thereby achieve energy savings.

Although local governments may play a small role in overall energy conservation, local circumstances do affect the effectiveness of federal and state programs. Growing areas present greater opportunities to affect energy consumption than do slowly growing or declining cities and towns. New schools and other public buildings can be designed for maximum energy efficiency. Moreover policies to require energy-efficiency standards for new construction or to encourage residential growth in close proximity to employment obviously will have their greatest impact where private investments are great.

The effectiveness of any given policy will depend upon the particular circumstances of the urban area. For example, a policy to channel population growth to high-density central areas will reduce energy used in commuting if employment is centralized, but it could actually increase transportation energy use if employment is very decentralized. Policies to encourage retrofit of existing public or private buildings will generate much greater energy savings where heating and cooling requirements are greatest; the effectiveness of energy retrofit will also depend upon the existing building styles and construction types. Any federal or state policy to influence local government actions should therefore be sensitive to local variations of these sorts.

A modest role for local government in energy conservation should not be viewed as undesirable since local governments are not well suited to instituting energy-conservation policies that go beyond the individual self-interest of their residents. Local governments are likely to wait for others to make sacrifices to reduce energy use and decrease dependence on foreign sources. Policies that seriously conflict with individual self-interest therefore should be the responsibility of federal decision makers.

## Notes

1. Irving Hoch, *Energy Use in the United States by State and Region* (Baltimore: Resources for the Future, 1977).
2. E.A. Ide et al., *Estimating Land and Floor Area Implicit in Employment Projections: How Land and Floor Area Usage Rates Vary by Industry and Site Factors,* NTIS PB-200-069 (Washington, D.C.: Government Printing Office, July 1970).
3. Oak Ridge National Laboratory, *Commercial Demand for Energy: A Disaggregated Approach*, (Oak Ridge, Tenn., 1977).
4. U.S. Department of Commerce, *Energy Conservation Handbook,* reported in Bureau of National Affairs, *Energy Users Report* (Washington, D.C.: Bureau of National Affairs, 1976).
5. Oak Ridge National Laboratory, *Commercial Demand.*
6. Some evidence is available on what can be accomplished through comprehensive local action. For example, the city of Clearwater, Florida, estimates that it has reduced governmental energy consumption by 25 to 30 percent through building retrofit, improved management, and modified operating policies for its transportation fleet.
7. Energy and Environmental Analysis, *Energy Consumption Data Base, Household Sector Final Report* (Arlington, Va., April 1977), vol. 3.
8. Eric Hirst, "Residential Energy Use Alternatives: 1970 to 2000," *Science*, December 17, 1976, pp. 1247-1252.
9. Eric Hirst and Janet Carney, "Energy and Economic Effects of Residential Energy Conservation Programs" (Paper presented at the Twelfth Intersociety Energy Conversion Engineering Conference, August 1977).
10. Ibid.
11. U.S. Council of Environmental Quality, *The Good News about Energy* (Washington, D.C.: Government Printing Office, 1979).
12. Whether this translates into actual energy efficiency, of course, depends on use patterns. The calculation assumes that the total number of housing units increases by 2.2 percent per year between 1975 and 1985 and that pre-1970 units are removed from the stock at a rate of about 1 percent per year. No allowance is made for changes in the distribution of housing unit types.
13. House of Representatives, Subcommittee on Energy and Power, Testimony by Lawrence S. Mayer, May 22, 1978, 95th Congress, 2nd session.

14. Corbin Crews Harwood, *Using Land to Save Energy*, (Cambridge, Mass.: Ballinger, 1977).

15. Hittman Associates, *Residential Energy Consumption: Detailed Geographical Analysis*, prepared for U.S. Department of Housing and Urban Development (Washington, D.C.: Government Printing Office, 1972).

16. For example, see Jack E. Snell, Paul R. Achenbach, and Steven R. Peterson, "Energy Conservation in New Housing Design," *Science*, June 25, 1976, pp. 1305-1311.

17. Estimated from data in U.S. Census Bureau, *Annual Survey of Housing: 1975* (Washington, D.C.: Government Printing Office, 1976).

18. For example, one study in northern New York State found that winter energy usage could be reduced 5 to 10 percent through vegetation buffers and orientation. See "Energy Conservation Site Design Case Studies: Proceedings of the Second Workshop," Argonne National Laboratory Informal Report ANL/ICES-TM-22 (January 1979).

19. There is an extensive planning literature on this theme. See, for example, Richard F. Babcock, *The Zoning Game* (Madison: University of Wisconsin Press, 1966).

20. See Meta Systems, "Resource Use and Residuals Generation in Households," report prepared for U.S. Environmental Protection Agency, EPA 600/5-79-005, March 1979; John Weichelt, Testimony before the House Subcommittee on Energy and Power, January 26, 1977; and Federal Energy Administration data.

21. Massachusetts Energy Office, *Energy Conservation Analysis Program Final Evaluation Report* (Boston, Mass.: Massachusetts Energy Office, November 1978).

22. Congressional Budget Office, *Urban Transportation and Energy: The Potential Savings of Different Modes* (Washington, D.C.: Government Printing Office, September 1977).

23. M. Chang and A. Horowitz, "Estimates of Fuel Savings through Improved Traffic Flow in Seven U.S. Cities" (Detroit, Michigan: General Motors Research Laboratories, August 1978).

24. For example, see C.C. Harwood, *Using Land to Save Energy* (Cambridge, Mass.: Ballinger, 1977).

25. G.E. Peterson and W. Bateman, eds., *Urban Development Patterns* (Washington, D.C.: Urban Institute, forthcoming).

26. For a discussion of the influence of sanitation facilities on urban land use, see R. Tabors, M. Shapiro, and P. Rogers, *Land Use and the Pipe* (Lexington, Mass.: Lexington Books, D.C. Heath, 1976). For a discussion of the influences of transportation facilities on urban growth, see D. Harrison, "Transportation and the Dynamics of Urban Land Use" Discussion paper D78-22, (Cambridge, Mass.: Harvard University, December 1978).

27. For a compendium of growth control papers, see R.W. Scott, ed., *Management and Control of Growth* (Washington, D.C.: Urban Land Institute, 1975), vols. 1-3.

# 9 Institutional Constraints upon Environmentally Sound Energy Policy

*Gregory A. Daneke*

A recent national survey sponsored by the President's Council on Environmental Quality (CEQ) suggests that most American's are still fundamentally committed to enhancing their relationship to the natural environment.[1] This commitment, however, has been clouded by mounting economic pressures (increased by energy uncertainties) and a diminishing faith in government. "Deregulation" is once again the national hue and cry. Industrialists promise increased performance if government and its environmentalists would "get off their backs." Meanwhile, economists extoll the virtues of the "free market," while forgetting that the value of clean air and water continue to be underrepresented in the market place. The environment, unlike many other policy issues, does not merely go away because national priorities drift somewhat. But, maintaining environmentally sound policies in such an era presents an arduous task.

Nowhere, perhaps, is the task of maintaining environmentally sound programs and projects more difficult than in the area of energy policy. This difficulty stems initially from the fact that energy and environment are misperceived as some sort of dichotomy. Suggesting that they are not actually dichotomous does not imply, however, that they are not arch rivals. It merely suggests that along the continuum of natural resource utilization and technology development, wise and prudent usage dictates a fairly harmonious relationship between energy and the environment. Viewed fundamentally, divorcing energy from environmental considerations makes little or no sense, for energy is merely a means for achieving social ends, and a sound environment is one of those more essential ends. However, recognizing this symbiotic relationship and manifesting that recognition in the energy policy process are very different matters. Environmentally sound energy decisions are rendered unlikely by a number of institutional and organizational constraints. Some of the more profound constraints are a general policy climate that perpetuates a basic misunderstanding of the energy transition; a focus on procedural mechanisms and patterns of analysis that largely fails to embody environmental considerations in the actual decision process; and an organizational setting that places agencies with direct

Many of these ideas were gleaned while I served with the National Energy Planning Review Team/U.S. General Accounting Office. I would, therefore, like to thank my colleagues at GAO for numerous insights; however, they are not responsible for the opinions expressed here.

responsibility for maintaining environmental integrity in a position of relative impotence. These factors, of course, are not mutually exclusive, nor, perhaps, do they encompass the full range of issues associated with the apparent conflict between energy and environmental priorities. Nevertheless they provide a starting point for understanding where organizational reform might prove beneficial.

## Misunderstanding the Energy Problem

Energy is perhaps the most fundamental of all natural exchange processes. As Odum and Odum illustrate, all basic human and natural interactions can be understood in terms of energy flows.[2] Environmental degradation can be thought of as a manifestation of poor energy management in that it represents a failure to limit throughput or to balance material flows in such a way as to minimize entropy. The material well-being of a given civilization usually depends upon its ability to manage renewable resources wisely or constantly replenish its supply of nonrenewables. Since nonrenewables are by definition finite (or merely more difficult to obtain with time), a society based on nonrenewable energy resources will either have to invest ever-increasing amounts of capital to maintain its supply of a scarce resource or make a transition to other energy resources (or both). Either path is likely to have some disruptive effect upon the society's material well-being.

The United States currently is faced with just such a transition. How it chooses to deal with it will have profound implications for not only the material well-being but for all facets of the quality of life and particularly the natural environment. A gradual transition, emphasizing rigorous conservation measures and aimed at the development of renewable-energy resources, is likely to be the least disruptive of the economy and the environment.[3] More hasty measures, designed to satisfy short-range demands, on the other hand, may have serious negative consequences.

Understanding these potential difficulties and devising strategies to ameliorate them should be the purpose of energy-related environmental assessment. Unfortunately, there are few, if any, mechanisms in place to carry out this type of analysis. To the extent that the environmental impacts of a given energy development plan are known, there are virtually no procedures for integrating these considerations with basic technological and economic thinking. Environmental assessment currently is set up as an adversary process. Under such a process, energy analysts are primarily concerned with meeting procedural requirements.

### *Surviving the Energy Transition*

Whether one looks at the energy situation in terms of absolute scarcity, cartel-controlled production schedules, or the life expectancy of the Saudi royal family, the conclusion is the same: the era of cheap and dependable oil resources has ended. Assuming that this fact brings only higher oil prices and not capricious interruptions, it would not be too difficult for the United States to adjust, although some economic dislocations will occur. However, such an adjustment has been made more difficult by the fact that the United States has held its own oil prices below the world market level (basically the OPEC price). This strategy, designed to ease the burden of higher world prices upon the American economy, has been patently misguided. By holding oil prices down, conservation was discouraged; exploration and development of new domestic resources was curtailed; an artificial scarcity was thus created; and this scarcity forced a greater dependency upon foreign oil.[4] In short, by attempting to forestall an energy transition, the federal policy has made matters worse. A smooth transition has been made more tenuous. The United States has made transitions in the past (for example, from wood to coal and coal to oil); yet even when aided by governmental interventions, these transitions took twenty to thirty years. Furthermore these transitions were engendered, for the most part, by quasi-natural market forces, and primary sources were not necessarily subject to catastrophic interruptions.

It is this potential for catastrophic interruptions that now lends the overwhelming sense of urgency to the nation's energy decision making, and it is this urgency that poses the greatest threat to environmentally sound decisions. Some alternative energy futures offer a relatively harmonious relationship with the natural environment. But in the current atmosphere of haste, the quest for such alternatives has been truncated, and political forces impose a form of analytical myopia. Motivated by a number of factors, essentially economic stability and a nebulous concern for national security, the focus of national energy policy (to the extent that it can be called focused) has become the rapid reduction of dependence upon foreign oil. Of course, there has been a good deal of milling and shoving over how this reduction is to be achieved. A panoply of proposals and counterproposals has emerged since the first oil crisis of 1973. However, canvassing the various energy plans and policy statements, one is likely to identify the following recurring themes:[5]

1. Oil and natural-gas prices should be increased gradually to reflect actual replacement costs (the OPEC price).
2. The nation should shift as much and as soon as possible to the use of its abundant coal reserves.
3. Nuclear development should be accelerated.
4. Inexhaustible resource technologies should be brought into the marketplace as rapidly as is feasible.

In the midst of the gas shortages in 1979, the president added to this list a new emphasis on some old standbys, generically known as synthetic fuels (or simply synfuels).[6] Synfuels include heavy oils (such as oil shale and tar sands), unconventional natural gas, and coal liquefaction and gasification. For political reasons, this resurrection of synfuels has highlighted oil-shale and coal development, both of which have a number of unresolved environmental as well as economic issues associated with them.[7]

Thus despite all of the rhetoric about conservation and the president's endorsement of the solar policy review (with its 20 percent solar future), the predominant emphasis of current national policy remains supply expansion through centralized facilities.[8] This emphasis tends to ignore the fact that reducing demand and decentralizing supply through various renewable-energy technologies is the simpliest and cheapest way to come to grips with the present energy shortages. Efforts in these areas remain afterthoughts rather than focal points of policy, however.

### *The Energy Mobilization Board: A Symptom of the Present Pathology*

This ongoing misunderstanding of the energy situation is most profoundly manifest in the recently created Energy Mobilization Board (EMB), and it is the potential activities of this board that offer the greatest threat to environmentally sound energy policy. The EMB, created by an act of Congress, is designed principally to speed the development of large-scale, energy-producing facilities. While it remains to be seen exactly what powers the board will exercise, it is generally assumed that the EMB will be able to expedite and/or bypass state and local licensing and permit procedures, and in some cases waive NEPA requirements (such as the environmental impact statement) in pursuing such avenues as synfuel plants, with the underlying assumption that environmental restrictions are slowing up the energy transition.[9] This, of course, is imprudent. First, expanding facilities to produce expensive (both economically and environmentally) oil substitutes and nuclear energy has little to do with the energy problem. Second, evidence suggests that even if these expansions were viable, the fast-track process still would fail to deliver them at a scale and price range sufficient to affect the oil problem.[10] Finally, aside from a few rare cases such as the Sohio pipeline from California, there is little evidence to suggest that environmental reviews slow down the actual construction process. Most of the delays in recent years have been the result of labor disputes and general ineptitude in the energy-facilities construction industry.[11]

The EMB was and is merely a symbolic gesture. As one congressman explained in reference to the EMB, "Congress wanted to do something, even if it's wrong." This divestiture of policy prerogative, when faced with complex problems, is very characteristic of congressional decision making. Like Egor, Congress believes that it can control these monsters once they have aided the president (Dr. Frankenstein) in creating them. Moreover, like the poor, impotent hunchback, Congress gets a vicarious charge out of the monster's powerful meanderings, usually in the form of pork-barrel projects for their constituencies. But of course, the monster, usually ends up turning on and destroying its masters, and there is certainly this potential in the EMB. Meanwhile the American public must resort to vigilantism (blocking plants with their bodies or merely developing their own conservation and supply strategies) in order to challenge the monster and move beyond the symbolism to some sort of effective energy policy.

The EMB is definitely a slap in the face to public organizations responsible for energy planning, and its various proposed auxiliaries constitute an even more serious affront to administrative due process. For example, the proposed energy security corporation (ESC) would be established as a quasi-private-sector organization operating with public funds. Managers of this corporation would not be subject to the same conflict-of-interest restrictions or accountability mechanisms that now govern the public service. Ronald Moe of the Congressional Research Service observes in reference to the ESC, "As more and more public functions are assigned to mixed public/private enterprises, the administrative structure of the federal executive establishment is threatened with political and functional disintegration."[12]

In essence, the fact that Congress is seriously considering this type of amateur socialism exhibits a profound lack of trust not only in the federal bureaucracy but also in democratic government. Unless public organizations move purposefully to reestablish their representation of the public interest, as well as to cut waste and inefficiency, it is likely that their role in the policy process will be enormously diminished.

## The Problem with Present Procedures

The EMB is more than a statement about the inadequacy of current energy decision making; it is also an expression of frustration with the procedures of environmental assessment. National policy makers believe, incorrectly, that environmental safeguards are preventing them from developing energy-supply options. Interestingly enough, most of the existing environmental assessment procedures do not even safeguard the environment, let alone affect energy decisions. And for different reasons, environmental analysts may well share this general disenchantment with current procedural requirements. Thus the EMB could conceivably emerge as an opportunity to establish a new set of analytical mechanisms more conducive to environmentally sound energy policies.

*General Problems with Environmental Assessment*

The basic problems encountered when existing procedures of environmental analysis are applied to energy-related issues are not unlike those that plague many other policy areas. In theory NEPA has a number of far-reaching implications, but through its application it has been constrained to the formal process of developing an environmental impact statement (EIS). While slowing the decision process sufficiently to allow environmental forces to form coalitions and, in some cases, stop projects, the EIS has produced few of the results that the framers of NEPA intended.[13] In their empirical study of a range of projects, Hill and Ortolano concluded that the EIS process does not actually penetrate the basic values that guide resource decisions, nor does it seriously affect the selection of alternative courses of action.[14] With specific regard to unquantified environmental values, Andrews and Waits observed that "no agency . . . has explicitly implemented this charge."[15]

Beyond its failure to ensure that environmental concerns are given complete consideration, some observers contend that the EIS process actually obscures environmental values while bogging down resource decisions. Bardach and Pugliaresi suggest that the EIS process fails to take a "hard look" at environmental issues and cite the following specific drawbacks:[16]

1. It establishes an atmosphere of legal harassment, and thus encourages agencies to prepare a "defensive document," designed to merely avoid litigation.
2. The EIS itself has become an enormous tome, filled with highly technical minutiae and standardized observations, generally lacking in creative insight.
3. EIS writers are generally divorced from the planning process (including public feedback) and are discouraged from drawing conclusions. Thus, most EISs fail to identify the critical policy issues.

In sum, the EIS may be of little use as a practical policy tool.

Yet it is noteworthy that the new Council of Environmental Quality (CEQ) guidelines, which stipulate a shorter length and provide for preproject conferences, may ameliorate some of these difficulties. Moreover, the burgeoning field of environmental mediation has shown promise in terms of enhancing and extending the EIS process.[17]

In general, it may be fair to say that NEPA (the National Environmental Policy Act of 1969) has never been fully implemented. In particular, its provisions for a systematic review of a range of relevant alternatives have yet to be rigorously enforced. Moreover the hope that NEPA might eventually be strengthened through the courts has been dashed by recent Supreme Court rulings. For example, in the case of *Stryckers Bay Neighborhood Council* v. *Karlen*, (1980), the Supreme Court overturned a lower court decision to delay a HUD project because of its "serious environmental consequences." The Court, in effect, ruled that the environment cannot be accorded any greater weight than any other considerations. In this case, the consideration of mere time was judged to be more important than environmental disruptions. The Supreme Court went on to say that courts are firmly restricted to enforcing only the most narrow procedural requirements of NEPA.

*Environmental Assessment and Energy Policy*

In the energy area, the relationship to NEPA has always been somewhat strained. In fact, one of the first cases that established the universal EIS requirement stemmed from the Atomic Energy Commission's claim of exemption from NEPA provisions.[18] Natural antagonisms with the Environmental Protection Agency (EPA), the watchdog of NEPA, have been exacerbated by the fact that new air and water quality missions of EPA tended to constrain energy objectives. And once again, the urgency associated with energy objectives tended to promote an attitude of almost military expediency. Caught up in President Carter's call for the "moral equivalent to war," the Department of Energy's first director, James Schlesinger, remarked, in reference to the strategic petroleum reserve, that: "rarely in the past has a major national effort of this kind been undertaken without a decision to go to wartime methods. In World War II you might have run roughshod over environmental constraints and that might have been the preferable way."[19] On another occasion Schlesinger observed, "How grateful we should be that the Japanese chose to attack Pearl Harbor prior to the passage of the National Environmental Policy Act."[20]

Given these personality conflicts and territorial battles, the Department of Energy has developed a highly defensive mode of environmental analysis. In particular, it uses an elaborate data-gathering capability and a simplistic yet awesome computer-simulation model known as the SEAS to form an impenetrable wall between it and its environmental detractors.[21] The SEAS model produces a "residual count" for a given energy alternative, as well as an index of cleanup costs. These highly aggregated indexes are incorporated into beautifully designed, full-color brochures, known as environmental readiness documents.[22] Although impressive, these various documents actually say very little about the amount of environmental consideration that goes into critical energy decisions. Moreover a careful look at SEAS-generated indexes points up several analytical shortcomings. For example, residual counts do not speak to issues of environmental health and safety, nor do they provide a detailed index of life-quality impacts. And although it is regional in scope, the SEAS model does not deal with the distribution of environmental costs and benefits across regions and income groupings. Finally, on grounds other than the relative cleanup costs (for those few technologies where costs are readily available) the SEAS does not facilitate a comparison of alternatives.

In all fairness, however, the SEAS was not designed to provide detailed environmental impact assessments. Rather it is a mid-range planning tool, which might be used to highlight areas for more in-depth analysis. Unfortunately, like many of the planning devices in the Department of Energy, it is either underutilized or misused. Consider the example of the environmental analysis for NEP II, the national energy plan presented to Congress in the spring of 1979. In the fall of 1978, a set of public meetings was held across the country to identify social and environmental values. However, at these hearings, the public was not given energy alternatives to comment upon. Environmental analysts commissioned to work on the NEP II were equally perplexed, since they, too, had little idea of what NEP II was going to be like. Thus they proceeded to generate information regarding items that they believed, but had not been so instructed, would be likely candidates—for example, an oil swap with Japan. As months passed, environmental analysts were left in the dark as to what NEP II might entail, and it was decided that they would await the president's choices in order to conduct an EIS. Thus the president was allowed to pick his set of energy proposals without the aid of a detailed comparison of environmental impacts. As it turned out, the final list of proposals was actually presented to Congress several weeks before the environmental analysis was completed. That analysis merely entailed a SEAS printout and not a full-blown EIS as had been promised. While this type of back-filling operation may not necessarily be the norm, it is somewhat indicative of the limited state of the art in energy-related analysis.

## Organizational Incapacity

To capitalize on the current dissatisfaction with energy-related environmental assessment, effort should be directed at developing a serious planning process. Such a process, modeled after strategic planning in the private sector, would begin by raising fundamental issues about the social goals and end uses of energy and conclude with recommendations as to the least environmentally disruptive energy-development pattern. In this way, perhaps, environmental considerations might affect the choice of energy alternatives.[23]

However, despite support for this type of energy planning on Capitol Hill, in the courts, and even in the business sector, it is not likely to take hold in the relatively near future, given the general lack of organizational commitment.[24] This organizational incapacity is exhibited not only in the modes of analysis but also in the roles that environmental analysts play within the organization configuration. Roles, in turn, stem from general determinations of organizational missions and mandates and from patterns of leadership and authority. All of these factors culminate in energy-related environmental analysts' being relegated to a rather insignificant role in the administrative policy process.

This capacity problem is present both within the Department of Energy and within the various external agencies (essentially the Department of Interior and the Environmental Protection Agency) that have specific missions of maintaining environmental integrity. This observation tends to question the conventional wisdom that would have external review agencies assume greater responsibility in the absence of responsiveness by the lead agency.

### *External Review: The Problem of Overloaded Agendas*

Given the current structure of energy policy making, it almost goes without saying that external review agencies would have little effect upon policy formulation. Since environmental information has impact only when given early consideration in the energy planning process, and external agencies are rarely privy to that process (to the extent that a real planning process exists), their effectiveness is delimited at the outset. Moreover, even if these agencies were included in the policy process, they do not have the time, staff, or analytical capability to provide much input.

This last factor is closely associated with what Richard Andrews has identified as the "problem of overloaded agendas." He explains how new energy-environment issues confront already-overloaded natural-resource agencies (including the Department of Energy) with more-diverse and often conflicting missions. To cope with this overload, an agency usually follows one of the following paths: spreading itself thin; meeting specific requirements first (such as the EIS); providing token responses; following the path of least resistance (social or political feasibility); choosing tasks on the basis of personal preference; or allowing the agency chief to set programmatic priorities.[25] In short, agencies that are not specifically oriented toward energy policy usually intervene very late and superficially in energy decision making. At this late juncture their role is usually adversarial.

Consider, for example the Environmental Protection Agency. Its relatively small energy staff is seldom asked to review Department of Energy plans until the process is completed, and at that point its criticism and recommendations are moot.[26] When involved in ongoing issues, the staff is inundated by complex and controversial calculations. In general, EPA has just too many other things to do to be much involved in energy. Its newest priority, hazardous wastes, only tangenitally linked to energy policy, is likely to absorb substantial agency resources for many years to come.[27]

The Council on Environmental Quality (CEQ), a tiny and relatively powerless executive advisory body, on the other hand, has made energy a major focal point. In fact, CEQ has spent so much time and effort arguing for an environmentally sound energy future that it has blown its cover and its credibility. This advocacy posture, accompanied by a staff insufficient in providing sophisticated analytical support, has made CEQ, at best, a resident gadfly.[28]

Energy analysts within the Department of Interior suffer from a similar lack of credibility, but from the other side of the environmental coin. Until very recently, Interior had never thought of itself as an environmental protection agency, and this ethos is still quite alien to it. Interior's administration of the public lands has usually emphasized an odd mixture of preservation and maximum utilization. Its present reluctance to grant energy-development leases may be as much a result of bureaucrat ineptitude as it is a conscious effort to respect environmental integrity.[29] This environmental ambivalence, coupled with limited programmatic purview, makes Interior a silent partner in energy policy.

*Internal Review: Analysis in Search of Policy Making*

Unlike the various external review agencies, the Department of Energy has ample analytical expertise. While several difficult issues (such as health and safety effects) escape causal determination or otherwise ellude exact measurement, DOE's Office of the Assistant Secretary for the Environment (EV) is at least equipped to deal with most environmental inquiries and is fully capable of comprehensive planning in the area of energy-environment interface. Yet it is rarely given the opportunity to engage in such planning, for this would, by definition, require a comparison of the relative environmental impacts of a full range of energy alternatives–for example, from conservation and solar applications to coal and nuclear. Such comparisons are mandated by NEPA, but those provisions have never been given effective enforcement in the courts and thus are rarely considered vital to NEPA compliance. Moreover such comparisons would be difficult, at best, in the current Department of Energy milieu, in which clear lines of demarcation are drawn between the multiplicity of research and hardware development projects, and where each little tub sits on its own bottom. Comprehensive cross-cutting comparisons are virtually impossible since diverse programs rarely communicate with one another or to any sort of central planning mechanism.

To the extent that coordination and establishing priorities, occur from time to time, EV's involvement in those processes is practically nonexistent. A recent Government Accounting Office report concluded that "although the Office of Environment has made the Department of Energy more aware of the need to consider environmental factors in developing energy technologies, it does not have a large role in the decision-making process which results in the selection and development of energy programs and projects."[30]

In response to these types of criticism and with a new secretary of energy, EV has made and will continue to make great strides in the area of basic NEPA compliance.[31] Moreover EV has recently gained some inroads into the budgetary and contract-letting processes which are tantamount to involvement in establishing priorities.[32] However, an authentic process for injecting social and environmental values into energy decision making still awaits the establishment of a variety of integrated impact assessment mechanisms, and the provisions for comprehensive strategic planning.

To initiate such a planning process, the Department of Energy might do well to pursue the following EPA recommendations:

1. Its project review process should be modified to provide a more-balanced and comprehensive review that includes more evaluation of such programs as energy conservation.
2. It should develop explicit environmental criteria for use in formulating energy policy and evaluating technology development, and submit those criteria for public and peer review. The criteria should facilitate quantitative cross-technology comparisons, where possible, and include explicit examination of health, environmental, and socioeconomic impacts.
3. It should systematically involve the public in a timely and meaningful manner in consideration of the environmental aspects of policy development, program planning, and project management.
4. It should clarify the role of the Environmental Advisory Committee and link the committee's functions more directly to departmental activities.[33]

These changes may not guarantee that environmental values will guide energy decisions, but they might provide a start toward environmentally sound energy policy.

## Conclusions

If the use of energy is merely a means of pursuing life-quality objectives rather than some sort of end in and of itself, then energy policy should represent broader social goals. Among such goals environment integrity would still seem to rank quite high, particularly if citizens are aware of the vital relationships between environmental integrity and health and economic well-being. At present, however, the nation seems bent on general energy situations that are likely to raise serious problems in terms of environmental integrity. Therefore it is incumbent upon environmental analysts and administrators to bring these fundamental incongruities to the attention of administrative as well as legislative policy makers.

This, however, is difficult, at best, under current institutional constraints. External mechanisms are constrained both by the timing and the authority of their inputs, and internal mechanisms are constrained by their lack of access to the process of establishing priorities. These factors need not prevail, however.

To the extent that planners and analysts turn their analytical capabilities inward and investigate their role within complex organizations, avenues of access may become apparent.

## Notes

1. Council on Environmental Quality, *Public Opinions on Environmental Issues* (Washington, D.C.: Government Printing Office, 1980).

2. Howard T. Odum and Elisabeth Odum, *The Energy Basis of Man and Nature* (New York: McGraw-Hill, 1976).

3. Council on Environmental Quality, *The Good News on Energy* (Washington, D.C.: Council on Environmental Quality, 1979); also see Gregory A. Daneke, "The Poverty of Energy Administration and Policy," in Gregory A. Daneke and George Lagassa, eds., *Energy Policy and Public Administration* (Lexington, Mass.: Lexington Books, D.C. Heath, 1980), pp. 9-29.

4. These results are explained in depth in Edward J. Mitchell, *U.S. Energy Policy: A Primer* (Washington, D.C.: American Enterprise Institute, 1974), esp. pp. 1-16.

5. See Executive Office of the President, *The National Energy Plan* (Washington, D.C.: Government Printing Office, April 1977); Department of Energy, *The National Energy Act* (Washington, D.C.: Government Printing Office, November 1978); DOE, *A Preliminary Outline and Approach: National Energy Plan II* (Washington, D.C.: Government Printing Office, October 1978); DOE, *Public Meetings on NEP II* (Washington, D.C.: Government Printing Office, October 1979); Office of Technology Assessment, *Analysis of the Proposed National Energy Plan* (Washington, D.C.: Government Printing Office, August 1977); DOE, *National Energy Plan II* (Washington, D.C.: Government Printing Office, May 1979).

6. Executive Office of the President, *Presidential Energy Initiatives* (Washington, D.C.: White House, Fact Sheet, July 16, 1979).

7. Charles A. Stokes, "Synfuels at the Crossroads" *Technology Review* 81 (August-September 1979):24-25; also see U.S. DOE, *An Environmental Analysis of Synthetic Liquid Fuels,* (Washington, D.C., July 12, 1979); and Government Accounting Office, "Arguments against Synfuels," memorandum (July 20, 1979).

8. For a more elaborate discussion of this distinction see Andy Lawrence and Gregory A. Daneke, "Issues Affecting Decentralization of Energy Supply," in Daneke and Lagassa, *Energy Policy,* pp. 63-79.

9. For discussions of the powers of the EMB, see Gene Coan and Carl Pope, "Energy 1979–What Happened and Why," *Sierra* 65 (January-February 1980):11-47; also see: Dirschten, "Cutting Energy Red Tape Raises Legal,

Practical Questions," *Natural Journal* 9 (1979): 1448-1451, and "Senate Approves Powerful New Energy Board," *Congressional Quarterly*, October 6, 1979, pp. 2187-2189.

10. See RAND Corporation, *A Review of Cost Estimation in New Technologies* (Santa Monica, Calif.: RAND, 1979).

11. See National Power Commission, *National Power Survey* (Washington, D.C.: FPC, 1974), pp. 79-80; also see Linda Cohen, "The Development and Utilization of Atomic Energy" (Paper presented to the Research Conference on Public Policy and Management, Chicago, October 1979).

12. Ronald Moe, "Government Corporations and the Erosion of Accountability: The Case of the Proposed Energy Security Corporation," *Public Administration Review* (November-December, 1979).

13. See Lynton K. Caldwell, "The Environmental Impact Statement: A Tool Misused," mimeo. (Public and Environmental Affairs, Indiana University, 1977).

14. William Hill and Leonard Ortolano, "NEPA's Effect on the Consideration of Alternatives: A Crucial Test" (NSF Report, Draft, Program in Infrastructure Engineering, Stanford University, 1978).

15. Richard N.L. Andrews and Mary Jo Waits, *Environmental Values in Public Decisions* (Ann Arbor, Mich.: School of Natural Resources, 1978).

16. Eugena Bardach and Lucian Pugliaresi, "Environmental-Impact vs. the Real World, *Public Interest* 49 (Fall 1977): 22-38.

17. See Roger Richman et al. "Environmental Mediation and the Administrative Process" (Paper presented to the Meeting of the American Society for Public Administration, Baltimore, April 1979); Larry E. Susskind, "The Use of Negotiation and Mediation in Environmental Impact Assessment" (Paper presented to the Conference of the American Association for the Advancement of Science, San Francisco, January 1980); also see John Busterud et al., *Materials on Resolve: Center for Environmental Conflict Resolution*, (Washington, D.C.: Resolve, 1979).

18. This is the case involving a nuclear plant at Calvert Cliffs, Maryland, and can be found in any text on environmental law.

19. From Garrett Epps, "The Creation of Energy," *Washington Post Magazine,* May 20, 1979, p. 19.

20. Ibid., p. 20.

21. SEAS was originally developed with EPA, which still runs its version of the model. For a laymen's view of the SEAS process, see MITRE Corp., *A Short Course on the SEAS Model* (McLean, Va.: Metrik Division, 1978).

22. Note, for example, DOE, Office of Technology Assessment, *EDR on Coal Gasification, Commercialization Phase 3 Planning* (Washington, D.C.: DOE, Assistant Secretary for Environment, August 24, 1978).

23. For an elaboration, see Gregory A. Daneke, "Strategic Energy

Planning: Toward an Integrated Impacts Approach" (Paper presented to the meeting of the American Association for the Advancement of Science, San Francisco, January 1980).

24. See George Brown (Address before the Conference of the American Planning Association, Baltimore, October 15, 1979); Richard Brooks, "The Legalization of Planning within the Growth of the Administrative State," *Administrative Law Review* 31 (Winter 1979); also note the classic, Charles Reich, "The Law of the Planned Society," *Yale Law Journal*, 75 (1966); Richard Heckart, "Environmental Alternatives and Social Goals," *Annals* 444 (July 1979).

25. Richard N.L. Andrews, "Environment and Energy: Implications of Overloaded Agendas," *Natural Resources Journal* 19 (July 1979):497-499.

26. Martin Wagner, "Comments on the Environmental Impacts of NEP II," memorandum to DOE from EPA (1979).

27. Thomas H. Maugh II, "Justice, EPA Begins Hazardous Wastes Drive," *Science* 207 (January 1980):162.

28. CEQ., *The Good News*.

29. General Accounting Office, *Federal Coal Leasing: Issues Facing a New System* (Washington, D.C.: Government Printing Office, September 1979).

30. General Accounting Office, *The Energy Department's Office of Environment Does Not Have a Large Role in Decision-Making,* Report to the Senate Committee on Governmental Affairs (Washington, D.C.: Government Printing Office, 1980).

31. Office of the Environment, *New Guidelines for Environmental Review* (Washington, D.C., U.S. DOE, 1980). (Note: not exact title, document was unavailable at this writing).

32. Drawn from discussions with DOE executive personnel, April 1980.

33. EPA, *The Federal Non-nuclear Energy Act: Environmental Evaluation,* EPA 600 19-80-008 (Washington, D.C.: Office of Environmental Engineering, 1980), pp. 2-3.

# Index

# About the Contributors

**Karen Burstein** is an attorney and was a member of the New York State Public Service Commission from 1978 to 1980. She was a member of the New York State Senate from 1972 to 1976. Dr. Burstein has served as lecturer and author in issues dealing with equal rights for women, nursing-home patients, utility regulation, and day care. She received the juris doctor degree from Fordham University Law School.

**Gregory A. Daneke** is associate professor and director of the environmental and energy planning and administration program in the School of Business and Public Administration, University of Arizona. He received the Ph.D. from the University of California at Santa Barbara and has taught planning and management at the University of Michigan and at the Virginia Polytechnic Institute, as well as serving as a visiting professor in the Center for Technology Administration at American University. He has been a member of the National Energy Planning Review Team of the U.S. General Accounting Office and a visiting scholar with the Energy Information Exchange Program of the U.S. State Department. He has published numerous works in the area of policy analysis, natural-resources management, and energy-systems assessment. Dr. Daneke has directed several research projects in the area of life-quality accounting and socioeconomic impact and is currently associated with Resource and Development Consultants International of Ann Arbor, Michigan.

**David Harrison, Jr.**, is an associate professor of city and regional planning at Harvard University. He has concentrated his research in environmental policy and land use. The author of *Who Pays for Clean Air*, he has recently completed a major study of the influence of transportation, income, and demographic trends on land use. In addition to urban energy policy, his current research includes a comparison of alternative environmental regulation approaches, focusing on the issue of airport noise. On leave during 1979-1980, he is serving as a senior economist on the staff of the President's Council of Economic Advisers.

**Helen Ingram** is professor of political science at the University of Arizona. She is a Fellow at Resources for the Future. Professor Ingram has published widely in the fields of natural resources and the environment and is the co-author of the recently published book, *A Policy Approach to Political Represenation: Lessons from the Four Corner States.*

**Michael E. Kraft** received the Ph.D. from Yale University and is an associate professor of political science and environmental administration at the University of Wisconsin at Green Bay. His research interests include environmental politics in the United States, public-policy analysis, population policy, and legislative behavior. He is the coeditor of *Population Policy and Analysis: Issues in American Politics* (1978), editor of *Environmental Policy Studies Directory*, and contributing author to *Population and Politics* (1973), *Environmental Politics* (1974), *The Sustainable Society* (1977), *Medical Ethics and the Law* (1980), and *Encyclopedia of Policy Studies* (forthcoming). His articles have appeared in the *Policy Studies Journal, Polity,* and *Alternatives.*

**Nancy Laney** received the masters of public administration from the University of Arizona, where she is currently pursuing her doctorate. Formerly employed by the Arizona State Senate, she is a member of the Governor's Commission on the Arizona Environment. Ms. Laney is coauthor of *A Policy to Political Representation: Lessons from the Four Corner States.*

**Walter A. Rosenbaum** received the Ph.D. from Princeton University and is a professor of political science at the University of Florida. His research interests include energy and environmental policy, public participation in administration, and the presidency. During 1978-1979, he was a Fellow at the Woodrow Wilson Center, Smithsonian Institution. His book on the politics of energy policy will be published in 1981.

**Dankwart A. Rustow** is Distinguished Professor of Political Science, City University of New York. He received the Ph.D. from Yale in 1951, has taught at Princeton and Columbia, and held visiting appointments at Harvard, Yale, Heidelberg, Torino, and the London School of Economics. He is a member of the Council on Foreign Relations and the International Association of Energy Economists. He has also served as vice president of the American Political Science Association (1973-1974) and the Middle East Studies Association of North America (1969-1970). Dr. Rustow is the author of *OPEC: Success and Prospects* (1976), *Middle Eastern Political Systems* (1971), *Philosophers and Kings: Studies in Leadership* (1970), *A World of Nations* (1967), and numerous articles and monographs in scholarly and general publications in the United States and abroad.

**Michael H. Shapiro** is an assistant professor of city and regional planning at Harvard University. He is interested in environmental and energy technology and in quantitative techniques of policy analysis. His research includes studies of the relationships between land use and environmental quality. Currently he is examining urban energy issues, including the application of solar technologies in existing urban areas. The coauthor of two books on sewerage planning, Professor Shapiro has published several papers on environmental-impact assessment and environmental-planning models.

**Lettie McSpadden Wenner** is associate professor of political science at the University of Illinois at Chicago Circle, where she teaches public law and public policy, with particular emphasis on environmental policies. She has published articles on the federal Clean Water Act, implementation of state water-pollution-control laws, the difficulties inherent in the use of user charges and cost-benefit analysis for the solution of various environmental problems, and public participation in decision making through referenda on the nuclear-power issue. Her text, *One Environment under Law* (1976) analyzes a variety of environmental problems and public policies designed to ameliorate those problems. Currently she is working on a book describing the role of federal courts in reviewing and overseeing environmental policies during the 1970s.

## About the Editor

**Regina S. Axelrod** is chairperson of the Department of Political Studies and is on the faculty of the Institute for Suburban Studies at Adelphi University. She has specialized in environmental and energy policy and decentralization of urban government. Currently, Dr. Axelrod is investigating the environmental impact of massive coal-conversion in the New York City metropolitan region. Dr. Axelrod directed the Conference on Energy and Environment, June 8-9, 1979, at Adelphi University. She received the Ph.D. from the Graduate School of the City University of New York in 1978.